Nonlinear theory of elastic stability

Monographs and textbooks on mechanics of solids and fluids

editor-in-chief: G. Æ Oravas

Mechanics of elastic stability

editor: H. Leipholz

1. H. LEIPHOLZ
   Theory of elasticity

2. L. LIBRESCU
   The elasto-statics and kinetics of anisotropic and heterogeneous shell-type structures

3. C. L. DYM
   Stability theory and its applications to structural mechanics

4. K. HUSEYIN
   Nonlinear theory of elastic stability

Nonlinear theory of elastic stability

K. Huseyin

University of Waterloo
Faculty of Engineering
Department of Systems Design
Waterloo, Ontario

NOORDHOFF INTERNATIONAL PUBLISHING
LEYDEN

© 1975 Noordhoff International Publishing
A division of A. W. Sijthoff International Publishing Company B.V.,
Leyden, The Netherlands

All rights reserved. No part of this publication may be reproduced, stored in a retrieval system, or transmitted, in any form or by any means, electronic, mechanical, photocopying, recording or otherwise, without the prior permission of the copyright owner.

ISBN 90 286 0344 1

Set-in-type in the United Kingdom
Printed in The Netherlands

*To
Tuncay, Neyzar, Zuhal and Hulya*

Contents

Preface xi

Introduction. xv

   (i) Introductory remarks xv
   (ii) Classification of systems xviii
   (iii) Concept of stability and energy criterion xix

1. One-parameter systems 1

   1.1 *Simple (discrete) critical points.* 3

      1.1.1 Basic concepts and definitions 3
      1.1.2 Energy expansion and equilibrium paths 6
      1.1.3 Perturbation methods 14
      1.1.4 Stability of an equilibrium state 17
      1.1.5 Critical loads 20
      1.1.6 Post-critical behaviour 25
      1.1.7 Stability distribution on the equilibrium paths in the vicinity of a bifurcation point 31

   1.2 *Imperfect systems: simple critical points* 35

      1.2.1 General remarks 35
      1.2.2 Equilibrium analysis 35
      1.2.3 Stability analysis: critical load-imperfection relationship 41

   1.3 *Symmetric systems: simple critical points* 47

      1.3.1 General remarks 47
      1.3.2 Potential energy and symmetry 48
      1.3.3 Equilibrium analysis 51

Contents

    1.3.4   Stability analysis: critical loads   53
    1.3.5   Post-critical behaviour and imperfection sensitivity .   56

1.4 Examples: simple critical points   61

    1.4.1   General remarks   61
    1.4.2   Limit points: one-degree-of-freedom model   61
    1.4.3   Asymmetric points of bifurcation: one-degree-of-freedom model   64
    1.4.4   Stable-symmetric points of bifurcation: two-degree-of-freedom model   67
    1.4.5   Shallow arch: unstable-symmetric points of bifurcation and limit points   70

1.5 Coincident critical points   79

    1.5.1   General remarks   79
    1.5.2   Stability of coincident critical points   80
    1.5.3   Equilibrium paths emanating from a coincident critical point: compound branching   86
    1.5.4   Symmetric systems   89
    1.5.5   Imperfect systems   93
    1.5.6   Illustrative example   101

2. Multiple-parameter systems   107

2.1 Fundamentals   109

    2.1.1   General remarks   109
    2.1.2   Basic concepts and definitions   110
    2.1.3   Classification of critical points   114

2.2 General critical points   123

    2.2.1   Estimation of the stability boundary   123
    2.2.2   Equilibrium surface in the vicinity of a general critical point   127
    2.2.3   Critical zone and the stability boundary   132
    2.2.4   The critical surface as the existence boundary . . .   136
    2.2.5   One-degree-of-freedom systems   138
    2.2.6   Singular critical points   141
    2.2.7   Stability distribution on the equilibrium surface . .   145
    2.2.8   Examples, experimental results and discussion . . .   148

Contents

2.3 *Special critical points: asymmetric* 159

    2.3.1 Equilibrium surface in the vicinity of a special critical point. 159
    2.3.2 Stability boundary. 164
    2.3.3 Stability distribution on the equilibrium surfaces . . . 166
    2.3.4 Linear eigenvalue problems 168

2.4 *Symmetric systems* 171

    2.4.1 Estimation of the stability boundary 171
    2.4.2 Stability boundary of a system with limited degrees of freedom 175
    2.4.3 Convexity of the stability boundary of systems associated with a fundamental equilibrium plane. . . . 176
    2.4.4 Equilibrium surface in the vicinity of a symmetric special critical point. 181
    2.4.5 Stability boundary in the vicinity of a symmetric special critical point. 188
    2.4.6 Stability distribution on the equilibrium surfaces . . 188

2.5 *Examples and experimental results* 195

    2.5.1 Four-degree-of-freedom model: symmetric special critical point 195
    2.5.2 Shallow arch under two independent sets of concentrated loads 198
    2.5.3 Shallow arch under uniform pressure and concentrated load at apex 200
    2.5.4 Experimental results. 206

Concluding remarks. 209

References 211

Index 217

Preface

Some form of instability is not an uncommon event in various phases of daily life providing, of course, one observes and identifies the situations. Against such a broad spectrum, this book is concerned with a rather limited though well-defined set of instability problems. A general nonlinear theory of *elastic* stability for *discrete conservative* systems is presented. Immediate applications of the theory are in the field of structural engineering from which the illustrative examples are all chosen. Nevertheless, the general formulation allows for application of the theory to problems in other engineering disciplines providing the essential variables and parameters are interpreted appropriately.

The theory treated in this book falls into three main categories:

1. It provides a phenomenological exposition of various aspects of instability behaviour associated with conservative systems. Basic nonlinear concepts of elastic stability are developed with regard to well-defined classes of systems.
2. Under some circumstances, the phenomenological studies lead to certain fundamental theorems which, in addition to providing for a general conclusive statement, may also prove to have an extremely important practical value. The theorems concerning the convexity of the stability boundary, for example, can yield upper and/or lower bound estimates of the boundary without much analytical effort.
3. The general theory also provides for systematic methods of nonlinear equilibrium and stability analysis, and in this exposition, *the multipleparameter perturbation technique* assumes a dominant role.

It is the author's belief that a newcomer to the field must first have a thorough understanding of the basic concepts and principles underlying stability theory before attempting to blindly tackle specific problems with incomplete and often mistaken notions. Such an overall view and insight can best be provided through a general theory which delineates various instability phenomena systematically. In particular, a theory which is

Preface

developed in terms of generalized coordinates can be very helpful in this regard because of its relative simplicity, and in fact, this is one of the main objectives of the present book. Graduate students in the departments of structural, civil, mechanical, aeronautical engineering as well as in applied mathematics and mechanics should, therefore, be interested in the material covered herein, and the book is very well suited for adoption as a text at the graduate level. The problems added to the end of appropriate sections are intended as exercises for students. Research workers will find interesting grounds to explore particularly in Part 2, and the methods of nonlinear analysis presented in various sections may prove very useful for engineers and analysts in general.

The introduction to the treatise includes historical notes, a classification of conservative systems and a brief discussion concerning the definition of stability and the energy criterion. The classified systems are, then, treated in two parts, Part 2 containing most of the material in Part 1 as a special case. This line of presentation is taken in order to increase the accessibility of classical one-parameter stability problems in the self-contained Part 1, in which a wide variety of instability phenomena are elucidated with regard to coincident as well as simple critical points together with the associated imperfection-sensitivity questions. Symmetric systems receive special attention in Chapter 1.3, and some of the results are given in explicit formula-type forms suitable for direct use perhaps with the aid of a digital computer.

Part 2 encompasses relatively recent developments, and it starts by laying down new definitions and introducing an appropriate reclassification of critical conditions (Chapter 2.1). The subsequent chapters expand the theory in two main directions identified in Chapter 2.1, the analyses being primarily based on the multiple-parameter perturbation technique. In such a broader characterization of instability behaviour, one acquires a deeper insight into the problems associated with one-parameter as well as multiple-parameter systems. In fact, through the perspective provided by the multiple-parameter analysis, a number of instability phenomena related to classical branching conditions are brought to light. The introduction of certain appropriate transformations leads to several fundamental theorems which have a practical as well as a theoretical value. Symmetric systems are again treated separately and fully.

Examples in both parts are chosen specifically for illustration of various concepts, methods and phenomena, and the results of a few experiments supporting the theory are presented.

I would like to express my gratitude to Dr. H. Leipholz, who inspired me to write this book. Thanks are also due to Dr. S. T. Ariaratnam and Dr. R. H. Plaut for reading the manuscript and making several valuable

Preface

suggestions. I am indebted to Dr. S. R. Parimi who carefully checked the solutions of examples and indicated corrections. Finally, I would like to thank Ms. R. Taylor, for typing the manuscript with amazing skill, and acknowledge the support of the National Research Council of Canada.

University of Waterloo,　　　　　　　　　　　　　　　Koncay Huseyin
December, 1974.

Introduction

(i) Introductory remarks

As Koiter says, 'the problem of elastic stability belongs inherently to the domain of the *nonlinear* theory of elasticity' although satisfactory results may be obtained from a linearized analysis in many important practical cases. Kirchhoff's uniqueness theorem which has a central role in the classical linear theory of elasticity is no longer valid. In fact, the simplest form of a loss of stability manifests itself in the appearance of more than one solution. In other words, many structures which have a dimension much smaller than the others—such as bars, thin plates and shells—are liable to exhibit more than one equilibrium configuration under a given external loading, thus evading Kirchhoff's Law. Such structures can undergo a marked change in the character of their deformation that is not associated with a failure of the material but rather represents a loss of stability of the original equilibrium configuration. This sort of behaviour can no longer be described with sufficient accuracy by the linearized strain-displacement or stress-strain equations, so the theory of elastic stability emerges as an important branch of the nonlinear theory of elasticity and, in a broader sense, applied mechanics. In the sequel, the words *linear* and *nonlinear* may be used under different, though well-defined, circumstances, and it is important to bear in mind that while a procedure of analysis can be described as *linear*, the strain-displacement relationships, for example, are always *nonlinear* in structural instability problems. The principle of superposition and the principle of proportionality which are fundamental in linear mechanics are no longer valid.

Originating in Euler's classic investigation of an axially compressed column in 1744, the theory of elastic stability has become a very well-developed branch of Applied Mechanics. Most of the significant achievements, however, have been realized rather recently, two centuries after Euler's pioneering work. Despite the fact that Euler's theory includes a complete post-critical analysis, the problem of the theory of

Introduction

stability has long been considered to be the determination of the stability limit. This is partly because the post-critical region was, for a long time, of secondary practical interest and partly because the governing differential equations in this region are, in general, nonlinear, posing fundamental analytical difficulties.

The extensive use of plate and shell structures in the construction of aircraft, missiles and space vehicles, however, has forced the research workers to consider the post-buckling behaviour of structures with a view to meeting the increasing demands for economy in weight. As remarked before, while a linearized analysis may yield the critical load in many important practical problems, the behaviour in the post-critical range is essentially nonlinear. Similarly, the effect of imperfections on the buckling behaviour can only be studied through a nonlinear analysis. Furthermore, there are several important structural systems (such as shallow arch and shallow spherical cap) whose buckling behaviour including the determination of the stability limit requires a nonlinear analysis. Thus, while *ad hoc* approximate methods were being pursued in the solution of such specific nonlinear problems, the extremely complex nature of nonlinear buckling and post-buckling behaviour of shell structures in particular has stimulated a thorough re-examination of the basic concepts of elastic stability. As a result, the fundamentals of the stability theory have undergone a systematic development in the framework of a 'general nonlinear theory of elastic stability' in recent years.

The foundations of the general theory of elastic stability were laid by Poincaré (1885) in his classical paper on the stability of rotating liquid masses [1]. Using the concept of generalized coordinates, he studied the equilibrium path configurations in the vicinity of a critical equilibrium state and indicated the possibility of limit points, having devoted most of his nonlinear analysis, however, to branching conditions. He showed that a loss of stability is normally associated either with a *limit* point, which represents a local extremum on an initially stable path, or with a point of *bifurcation* at which the fundamental equilibrium path intersects a second distinct path, and an exchange of stabilities occurs. Poincaré's theory provided a valuable insight into the nature of the loss of stability, and although it was developed in terms of generalized coordinates for mechanical systems, it describes the overall buckling characteristics of a continuous elastic body as well. The systems considered by Poincaré and by other researchers later, however, have a single variable loading parameter, and it will be seen in Part 2 that multiple-parameter systems (with independent loads) can exhibit certain critical equilibrium states which appear as a limit point on one path of loading, and as a point of bifurcation on another. Such points are associated with *anticlastic*

(i) *Introductory remarks*

equilibrium surfaces (saddle surfaces) and will be called *general* while those associated with *degenerate* equilibrium surfaces resulting in bifurcation under all possible paths of loading will be called *special*. This reclassification of critical points will prove extremely useful in the general treatment of multiple-parameter systems.

In the context of continuum elasticity, Bryan (1888) appears to be the first researcher to attempt developing a general theory of stability [2]. He based his analysis on the *energy criterion* as a postulate generalized directly from the well-known Lagrange's Theorem for discrete mechanical systems, and it is believed that this is the first adaptation of the extremum properties of the potential energy to the continuous systems. His energy expression, however, contained quadratic terms only, and the scope of the analysis was, therefore, limited. The subsequent studies, mainly by Southwell [3], Biezeno and Hencky [4], Reissner [5], Trefftz [6, 7], Marguerre [8], Kappus [9], and Biot [10, 11], were aimed mostly at determining the stability limit rather than examining the behaviour of the system on reaching and exceeding this limit. Trefftz is credited with the first general formulation of the variational principle which yields the critical loads.

Koiter, in a thesis (1945) which is now a classic [12], presented a systematic nonlinear theory of stability and developed it further in several subsequent publications [13–18]. Koiter's theory, which is rigorous in an asymptotic sense, has contributed most significantly to our understanding of the nonlinear concepts of elastic stability. He continued Trefftz's analysis in the critical case when the second variation of energy fails to give sufficient information about stability, and derived the necessary and sufficient conditions of stability of a *simple* as well as a *coincident* critical point by resorting to higher order variations of energy as may be required. In his nonlinear post-buckling analysis, Koiter examined in detail the characteristic features of three discrete branching points which can be described as *asymmetric, stable symmetric* and *unstable symmetric*. The effect of small imperfections on the buckling behaviour was also studied. One of the most significant results of this refined general analysis is that the initial stage of post-critical behaviour is completely specified by the stability or instability of equilibrium at the critical point itself. His results explain why the difference between the critical loads of perfect and slightly imperfect structures may be considerable, and ties the *imperfection sensitivity* to the instability of the critical point itself. In his more recent work, Koiter examines the energy criterion, which is so widely used in the analysis of discrete mechanical systems as well as continuous systems, from a more critical point of view, and points out that even Lagrange's Theorem is, in fact, concerned with a *sufficient* condition only.

Introduction

To prove that a proper relative minimum of the energy is also a *necessary* condition, he introduces damping into the system, and later generalizes this approach to continuous elastic bodies.

Koiter's original work of 1945 attracted relatively little attention until the early 1960's while many investigators were preoccupied in an activity directed at obtaining more accurate results for thin shell structures and in particular for cylindrical shells under various loading conditions. Researchers in this field were, indeed, making an effort to explain the large discrepancies between the test results and the theory, following the work by von Karman and Tsien [19]. Interest in the general theory sprang up in the early 60's almost simultaneously at University College in England and at Harvard University in the United States. Budiansky, Hutchinson and their associates at Harvard continued developing a dynamic buckling theory of imperfection-sensitive structures parallel to Koiter's work and applying the latter to a variety of shell structures [20–26], while a second line of study in the general theory, using the concept of generalized coordinates, was undertaken by Chilver, Thompson and their associates at University College, London. The latter approach, being simpler mathematically, led to extensive developments of new physical concepts. In fact, some of the theories developed in this context have not yet been studied in the context of continuum elasticity (e.g., the material presented in Part 2). The major goal of this book is to present a systematic and comprehensive general theory of elastic stability based on the work of researchers originating from or associated with the University College.

(ii) **Classification of systems**

Problems of the theory of elastic stability are concerned mainly with two classes of systems, namely, conservative and nonconservative. The distinctive feature of the latter systems is that they involve external forces which do not possess a potential. It is apparent from the introductory remarks that this book is to be devoted to the study of elastic conservative systems. Gyroscopic conservative systems are also excluded.

Conservative systems have been classified in Table 1, which indicates various streams of activity in this field. Full lines circle the areas which will be treated in the sequel while dashed lines indicate those not to be considered. In addition, specific studies concerning the bifurcation properties of frames [27, 28], the behaviour of degenerating structures such as heavily-cracked cooling towers [29], the effect of random imperfections in struts [30], and perturbation techniques aimed at finite-element applications [31] will not be included. For a general discussion of

(iii) Concept of stability and energy criterion

Table 1

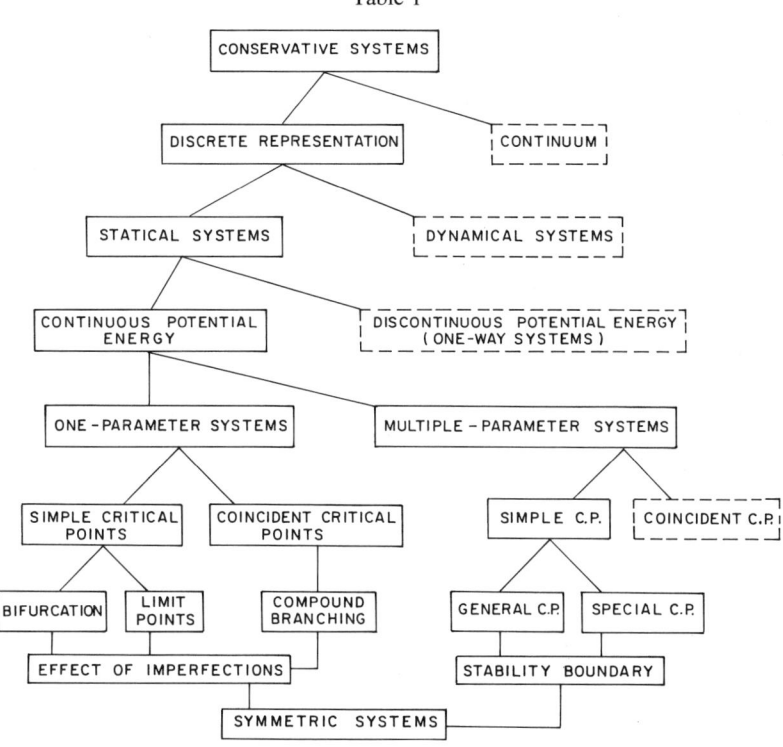

the equilibrium and stability of one-way structural systems, the reader is referred to the recent work of Burgess [32].

The general theory presented here is concerned with structural systems which are

(a) discrete
(b) deterministic
(c) elastic conservative
(d) statical in the sense of statical loading and statical behaviour

The analysis is mostly asymptotic.

(iii) Concept of stability and energy criterion

What is meant by stability? This question has no universal answer. A meaningful definition of stability can only be given within the realm of a broad field of applied sciences or, even more restrictively, in accordance with the specific requirements of a particular set of problems. In

Introduction

mechanics, Lyapunov's dynamical definition of stability for discrete mechanical systems stipulates that an equilibrium state is stable whenever in the motion following a sufficiently small initial disturbance of equilibrium (in terms of initial displacements and velocities), the local displacements and velocities remain as small as desired for all positive time.

A direct generalization of this definition to continuous bodies, however, might lead to discrepancies. Koiter observed that it is not physically meaningful to require that the displacements and velocities following a sufficiently small disturbance remain as small as desired all over the body. Nevertheless, it seems that the extension of Lyapunov's dynamical definition of stability to continuous bodies is possible provided certain appropriate modifications are introduced. Such modifications are concerned basically with the introduction of a prescribed *metric* and defining the stability with respect to this metric rather than local values. As a measure of deviation from a given equilibrium configuration, Koiter proposes the L_2-norms of the displacement and velocity fields in the form

$$\|\mathbf{u}\|^2 = \frac{1}{V} \int \mathbf{u} \cdot \mathbf{u} \, dv$$

and

$$\|\dot{\mathbf{u}}\|^2 = \frac{1}{M} \int \rho \dot{\mathbf{u}} \cdot \dot{\mathbf{u}} \, dv$$

respectively, where $\mathbf{u}$ describes the displacements from the equilibrium, $\dot{\mathbf{u}}$ is the time derivative and M and V are the mass and the volume of the body with density ρ.

If the fundamental equilibrium state is given by the coordinates of $\mathbf{x}$, an adjacent configuration may be described by $\mathbf{x}+\mathbf{u}$; the fundamental state is stable if one is able to find

$$\delta(\epsilon, \epsilon') > 0$$

and

$$\delta'(\epsilon, \epsilon') > 0$$

for any sufficiently small $\epsilon > 0$ and $\epsilon' > 0$, such that following the initial disturbance

$$\|\mathbf{u}_0\| < \delta$$
$$\|\dot{\mathbf{u}}_0\| < \delta'$$

one has

$$\|\mathbf{u}\| \leq \epsilon$$
$$\|\dot{\mathbf{u}}\| \leq \epsilon'$$

for all $t > 0$.

(iii) Concept of stability and energy criterion

A further restriction on the disturbances is that the energy supplied into the system must satisfy the inequality

$$\Delta E \leq \delta^*$$

where $\delta^*(\epsilon, \epsilon') > 0$.

Koiter's modified definition of stability reduces to Lyapunov's original definition in the case of a discrete system. Thus, one has a single dynamical definition of stability for both discrete and continuous systems which is most desirable.

Having agreed on such a definition, one might ask whether ascertaining this definition directly is the only correct way of ensuring stability in particular applications. Fortunately, this is not so, and there are, indeed, powerful *stability criteria* which can be used to ascertain stability according to Lyapunov's definition indirectly. Such criteria can be either 'dynamical' or 'statical'. Methods in the former category, being closer to the definition of stability, are more general and can be applied to conservative and nonconservative systems alike. Nevertheless, in the stability analysis of statical systems, one desires to employ statical methods despite the fact that the concept of stability itself is a dynamical one. Thus, the *energy criterion* emerges as a powerful statical criterion which is applicable to systems possessing a potential. In spite of this limitation that external forces must have a potential, the energy criterion has been used widely and successfully in the literature. This is, of course, partly due to the fact that the majority of practical structural systems and particularly those in civil engineering are often associated with conservative loads. Only relatively recent developments in aeronautical, contemporary mechanical and missile engineering increased the significance of nonconservative forces.

The energy criterion, together with the principle of virtual work, will form the basis of the entire analysis throughout this book. The well-known principle of virtual work states that *the total potential energy is stationary with respect to every kinematically admissible displacement from a state of equilibrium*. This is a necessary and sufficient condition. Lagrange's theorem, on the other hand, relates Lyapunov's dynamical definition of stability to the extremum properties of the potential energy. This theorem (its proof can be found in books on classical mechanics) ensures that *a state of equilibrium is stable if the potential energy has a proper relative minimum at that point*. This is a sufficient condition for stability, but whether a proper relative minimum is also a necessary condition apparently has not yet been solved in complete generality. Existing proofs concerning the instability of equilibrium in the absence of a proper minimum rely on certain additional assumptions such as the one

Introduction

given in the book by Chetayev [33]. As remarked earlier, Koiter [16] has tackled this problem by introducing damping into the system with a positive definite energy dissipation, and established a general theorem: '*An equilibrium configuration in which the potential energy has no proper relative minimum is always unstable in the presence of damping with a positive definite energy dissipation*'. Since some damping is always present in actual physical systems this proof is entirely adequate for all practical purposes. This result was later extended to continuous bodies [17].

Part 1

One-parameter systems

1.1

Simple (discrete) critical points

Actual solids and structures are never absolutely rigid and somehow or other deform under the influence of external loads. But in many cases these deformations are small and their effect on the conditions of equilibrium is negligible. Thus, a complete branch of mechanics treating finite-degree-of-freedom mechanical systems has been developed over centuries. On the other hand, many practical elastic solids and structures may deform significantly with an infinite number of degrees of freedom, but their overall configuration can still be described effectively by a finite set of coordinates for specific purposes. This can be achieved, for example, through application of Galerkin or Rayleigh–Ritz procedures. Finite-element technique is another way of representing a continuum by a finite number of degrees of freedom.

In the theory of elastic stability, the finite-degree-of-freedom discrete representation of a continuum problem, via the concept of generalized coordinates, leads to a rapid investigation of buckling as well as post-buckling characteristics. Exact continuum theories are almost inapplicable to practical situations, and the analysis of structural stability problems is often performed through Rayleigh–Ritz or other procedures which effectively discretize the system. The concept of generalized coordinates contributed significantly to our understanding and developing of new physical concepts in the context of the general theory of stability.

The general theory here as well as in Part 2 will be developed in terms of generalized coordinates, each coordinate representing one degree of freedom.

1.1.1 Basic concepts and definitions

A general elastic conservative system described by N generalized coordinates Q_i and a variable loading parameter Λ is considered. A given set of the Q_i ($i = 1, 2, \ldots, N$) defines the configuration of the system completely. In specific applications the loading parameter Λ will often

1.1 Simple (discrete) critical points

represent the magnitude of an external load, the proportional loading under several loads or even some other variables independent of the Q_i and influencing the behaviour of the system. For example, lengths and imposed deflections in rigid loading (independent of the Q_i) can be mentioned.

For each value of Λ, the system possesses a total potential energy consisting of the strain energy and the potential energy of applied loads. One can, then, define a total potential energy function

$$V = V(Q_i, \Lambda) \qquad (1.1.1)$$

which depends on the independent variables Q_i and Λ. It will be assumed here and elsewhere in this book that this function is single-valued and continuously differentiable at least in the region of interest.

In many structural problems the function (1.1.1) can be expressed as

$$V = U(Q_i) - \Lambda E(Q_i) \qquad (1.1.2)$$

where U is the strain energy and E the generalized displacement corresponding to Λ. This specialized system often exhibits features that are not valid for the general system of (1.1.1).

The necessary and sufficient condition of equilibrium, that the function V is stationary with respect to the Q_i (i.e., $\delta V = 0$), yields a set of N equilibrium equations

$$V_i(Q_i, \Lambda) = 0 \qquad i = 1, 2, \ldots, N \qquad (1.1.3)$$

where the subscript on V denotes partial differentiation with respect to the corresponding generalized coordinate. Similarly, partial differentiations with respect to Λ will be denoted by primes on V.

These equations define a series of *equilibrium paths* in the $(N+1)$ dimensional *load-deflection* space spanned by Λ and the Q_i which is also referred to as *configuration space*. A set of prescribed variables (Q_i, Λ) is represented by a point in this Euclidean space, and the correspondence between the points of the space and all sets of variables is unique. In particular, the points of equilibrium are identified as those satisfying the equation (1.1.3) and forming certain equilibrium paths. These paths might be composed of several self-curvilinear intersecting parts, and attention here is naturally focussed on their form and possible connection between stability and topological characteristics of such forms. As a matter of fact, such a connection seems most likely when one observes that the defining equations (1.1.3) of the equilibrium paths result from the vanishing of the first variation of the energy function while the concept of stability is linked to the second and higher order variations of the same potential function.

1.1.1 Basic concepts and definitions

It seems desirable, at this stage, that the concept of stability and the energy criterion as applied to the function (1.1.1) are discussed briefly. Consider an arbitrary state of equilibrium, F, lying on an equilibrium path defined by (1.1.3). Evidently, such a point in the load-deflection space, satisfying the necessary conditions, represents a potential extremum of the energy function

$$V = V(Q_i, \Lambda_F) \qquad (1.1.4)$$

at the constant loading level Λ_F, V being envisaged as a surface passing through F. A sufficient condition that V has a relative minimum at F is the positive definiteness of the second variation of the function (1.1.4), and by virtue of the energy criterion this condition becomes sufficient also for stability. If the second variation is negative definite, negative semi-definite or indefinite, i.e., if the quadratic form representing the second variation admits negative values, then, the possibility of a minimum and consequently stability is ruled out (see Table 2).

Table 2

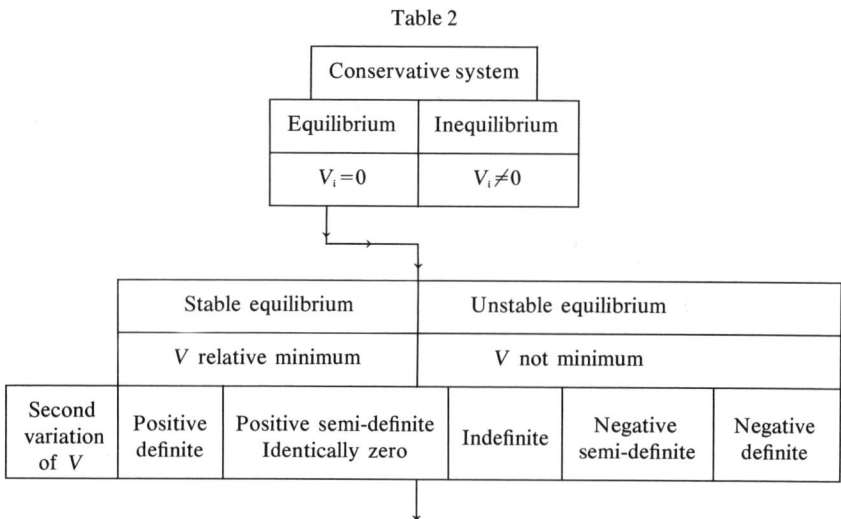

A remarkable situation arises when the second variation is positive semi-definite and the quadratic form admits zero values; such a state of equilibrium will be called *primary critical*. Obviously, the so-called *stability determinant* evaluated at such a point vanishes,

$$\det |V_{ij}| = 0 \qquad (1.1.5)$$

1.1 Simple (discrete) critical points

There are, of course, other equilibrium points which are not primary critical but, nevertheless, satisfy (1.1.5) and the associated quadratic form admits zero values; such equilibrium states at which the stability determinant vanishes will be called *critical*.

It follows that a critical equilibrium state is unstable unless it is primary in which case it can be stable or unstable depending on whether or not the potential energy function V has a relative minimum. As indicated in Table 2, in this case, the second variation fails to provide sufficient information concerning the maximum-minimum properties of the energy function, and higher order variations have to be examined. The vanishing of the third variation, then, becomes a necessary condition for stability, allowing for the possibility of a minimum in the fourth variation. The positive definiteness of the fourth variation is, then, a sufficient condition for stability, ensuring a relative minimum. A similar pattern of arguments follows in case the fourth variation is positive semi-definite.

A formal analytical procedure aimed at establishing the necessary and sufficient conditions for stability (essentially, conditions for a minimum of a function, $V(Q_i, \Lambda_F)$, of several variables, Q_i) will be presented in Section 1.1.4, and it is being deliberately deferred now in order to gain first an insight into the overall behaviour of the system as reflected in the load-deflection space, and to explore the methods to be applied in the sequel.

1.1.2 Energy expansion and equilibrium paths

The N equations of equilibrium (1.1.3) relate the $N+1$ independent variables Q_i, Λ. It follows that these relations have *one* free-ranging parameter and the equilibrium paths defined by (1.1.3) in the $N+1$ dimensional load-deflection space are, therefore, in the form of certain space curves.

An equilibrium path in the region of interest can be envisaged in the general parametric form

$$Q_i = Q_i(\eta), \quad i = 1, 2, \ldots, N$$
$$\Lambda = \Lambda(\eta) \tag{1.1.6}$$

where η is a real unspecified path parameter or, by choosing one of the basic variables (Λ say) as independent, in the form

$$Q_i = Q_i(\Lambda) \tag{1.1.7}$$

An equilibrium path may have stable and unstable parts, and one of

1.1.2 Energy expansion and equilibrium paths

the interesting questions appears to be concerned with the form taken by a path in the vicinity of a critical point.

The parametric representation (1.1.6) of the paths suggests that the neighbourhood of an equilibrium state can be explored analytically by means of a perturbation approach, and such methods will indeed form the basis of most of the analyses. In this section, however, an alternative method based on Taylor's expansion theorem will be introduced for an overall discussion of what may commonly occur in load-deflection space.

Consider an arbitrarily chosen equilibrium point, F, in the load-deflection space. The potential energy of the system will be referred to this *fundamental* state by writing it in the form

$$V = V(Q_i^F + q_i, \Lambda^F + \lambda) \qquad (1.1.8)$$

where q_i and λ are used to denote increments in the variables Q_i and Λ respectively. Initial interest is in the *fundamental equilibrium path* passing through the origin, and F can be envisaged on this path; the following analysis, however, is valid for an arbitrary F.

A linear orthogonal transformation

$$q_i = \alpha_{ij} u_j, \qquad \alpha_{ij}\alpha_{jk} = \delta_{ik} \qquad (1.1.9)$$

will further be introduced to diagonalize the quadratic form (in the q_i) of the energy expansion around F (Figure 1.1). Here and in the sequel

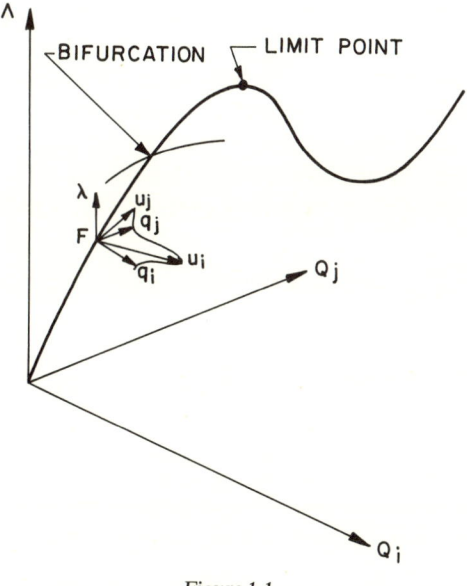

Figure 1.1

7

1.1 Simple (discrete) critical points

the dummy-suffix summation convention is adopted, and δ_{ik} is the Kronecker delta. By insisting that the transformation matrix α_{ij} is orthogonal, a unique transformation which changes the coordinates from the q_i to the u_i is, in general, ensured. The u_i are called the *principal generalized coordinates*, and it will be seen later that if F happens to be a coincident critical point, then, the principal directions are not unique.

Introducing the transformation (1.1.9) into (1.1.8), a new energy function,

$$H(u_i, \lambda) \equiv V(Q_i^F + \alpha_{ij}u_j, \Lambda^F + \lambda), \qquad (1.1.10)$$

is obtained, which can also be written as $H(u_i, \lambda) \equiv H_0(0, 0) + h(u_i, \lambda)$. The expansion of this function into Taylor's series, then yields

$$H = H_0 + H_i u_i + H'\lambda$$

$$+ \frac{1}{2!}(H_{ii}u_i^2 + 2H'_i u_i \lambda + H''\lambda^2)$$

$$+ \frac{1}{3!}(H_{ijk}u_i u_j u_k + 3H'_{ij}u_i u_j \lambda + 3H''_i u_i \lambda^2 + H'''\lambda^3)$$

$$+ \frac{1}{4!}(H_{ijkl}u_i u_j u_k u_l + 4H'_{ijk}u_i u_j u_k \lambda + 6H''_{ij}u_i u_j \lambda^2$$

$$+ 4H'''_i u_i \lambda^3 + H''''\lambda^4)$$

$$+ \frac{1}{5!}(\cdots) + \cdots \qquad (1.1.11)$$

Here subscripts and primes on H's denote partial differentiation with respect to the corresponding u_i and λ respectively, all derivatives being evaluated at F. It is further noted that $H_i = 0$ since F is a point of equilibrium and $H_{ij} = 0$ for $i \neq j$ due to (1.1.9).

A typical equilibrium equation, $\partial H/\partial u_r = 0$, can immediately be obtained as

$$\frac{\partial H}{\partial u_r} = H_{rr}u_r + H'_r \lambda$$

$$+ \frac{1}{2!}(H_{rij}u_i u_j + 2H'_{ri}u_i \lambda + H''_r \lambda^2)$$

$$+ \frac{1}{3!}(H_{rijk}u_i u_j u_k + 3H'_{rij}u_i u_j \lambda + 3H''_{ri}u_i \lambda^2 + H'''_r \lambda^3)$$

$$+ \frac{1}{4!}(\cdots) + \cdots = 0 \qquad (1.1.12)$$

It is understood that the equilibrium equations $\partial V/\partial q_i = 0$ are now

1.1.2 Energy expansion and equilibrium paths

replaced by $\partial H/\partial u_j = 0$ on the basis of

$$\frac{\partial H}{\partial u_j} = \frac{\partial V}{\partial q_i}\frac{\partial q_i}{\partial u_j} = \frac{\partial V}{\partial q_i}\alpha_{ij} = 0 \qquad (1.1.13)$$

which follows from the identity (1.1.10).

The tensorial character of the V_i, V_{ij}, etc., (at F) is worth noting here. Thus, writing (1.1.10) in the form

$$H(u_k, 0) \equiv V[Q_i^F + q_i(u_k), \Lambda^F + 0] \qquad (1.1.14)$$

and differentiating with respect to u_k yields

$$H_k = \frac{\partial q_i}{\partial u_k} V_i \qquad (1.1.15)$$

at F. One observes that the V_i transform according to the *covariant* law which is equivalent to the *contravariant* law when the transformation $q_i = \alpha_{ik} u_k$ is orthogonal [34]. It follows that the sets V_i and H_k are the representations of a covariant tensor of rank one in the q_i and u_k coordinate systems respectively. More generally, it can be shown that the sets of quantities V_{ij}, V_{ijk}, etc., are the representations of second, third, etc. order tensors in the q_i coordinate system. To this end, differentiate (1.1.14) once with respect to u_k ($k = 1, 2, \ldots, N$) and for a second time with respect to u_l ($l = 1, 2, \ldots, N$) to get, at F,

$$H_{kl} = \frac{\partial q_i}{\partial u_k}\frac{\partial q_j}{\partial u_l} V_{ij} \qquad (1.1.16)$$

which shows that the sets of V_{ij} and H_{kl} are the components of a second order covariant tensor in the q_i and u_k coordinate systems respectively. The third and higher order transformation laws similar to (1.1.16) follow successively from (1.1.14).

Since the transformation $q_i = \alpha_{ij} u_j$ is nonsingular and reversible, the quantities V_i, V_{ij}, etc. can be expressed in terms of H_k, H_{kl}, etc. respectively, and it can, therefore, be concluded that if all components of these first, second, etc. order tensors vanish in one coordinate system then they necessarily vanish in the other and, indeed, in all admissible coordinate systems. This is the well-known *invariant* property of tensors.

It is further observed that if the energy function is in the form (1.1.2), which is linear in Λ, then, the terms involving second and higher order derivatives of H with respect to Λ in (1.1.11) will be identically zero.

Proceeding now to explore the vicinity of F, suppose first that this point is noncritical. It appears, however, that one must first identify the critical and noncritical states of equilibrium in the context of the new formulation. A critical state of equilibrium was defined in the preceding

1.1 Simple (discrete) critical points

section as one which satisfies the determinantal equation (1.1.5). It follows from (1.1.16) that at F

$$\Delta_u = \det |\alpha_{ij}|^2 \cdot \Delta_q \qquad (1.1.17)$$

where

$$\Delta_u = \det |H_{ij}| \quad \text{and} \quad \Delta_q = \det |V_{ij}| \qquad (1.1.18)$$

The determinants Δ_q and Δ_u are, then, recognized as the invariant functions known as *relative scalars of weight 2* in tensor theory. Obviously, if Δ_q vanishes so does Δ_u necessarily.

Since the quadratic form of H is diagonalized, one has

$$\Delta_u = \prod_{i=1}^{N} H_{ii}, \qquad (1.1.19)$$

and at a critical point at least one of the coefficients H_{ii} vanishes. Evidently, the stability of F can also be discussed with respect to the H_{ii} conveniently, and because of this significant role played by the H_{ii}, they are referred to as the *stability coefficients* [1]. Thus, one observes that if all $H_{ii} > 0$, then the quadratic form of H is positive definite and F is stable. If any of the H_{ii} is negative, then, the quadratic form admits negative values and F is unstable. Also, F is a *simple (discrete)* critical point if only one stability coefficient vanishes, while it is a *coincident* critical point if two or more coefficients vanish simultaneously.

The system is normally stable at the unloaded state, and one would expect that the fundamental path generated by gradually increasing the loading parameter Λ from zero will remain stable for sufficiently small values of Λ. Suppose now that F is in this region; the fundamental path in its vicinity can readily be obtained by considering the equilibrium equations $\partial H/\partial u_i = 0$ which yields to a first approximation

$$H_{ii}u_i + H'_i\lambda = 0. \quad \text{(Note that there is no summation on } i.\text{)} \qquad (1.1.20)$$

(1.1.20) represents N first order equations in which second and higher order terms in the u_i and λ have not been retained due to the incremental character of these variables. If, however, some of the coefficients in (1.1.20) vanish, the equilibrium equations might have to be modified and additional appropriate terms retained.

Equations (1.1.20) define a straight line in the vicinity of F, and a *unique* incremental linearity between the u_i and λ [35]. In other words, there can be only one equilibrium path through a noncritical equilibrium state (Kirchhoff Law).

Let F now move gradually along the fundamental path until one of the stability coefficients, H_{11} say, vanishes. It is of course conceivable that

1.1.2 Energy expansion and equilibrium paths

more than one stability coefficient may vanish simultaneously; such *coincident* critical points, however, will be treated separately (in 1.5) and are excluded from the present discussion. The point on the path at which H_{11} vanishes represents, then, a simple *primary* critical state. Assuming that all the other energy coefficients are finite, the equilibrium equations take the form

$$\frac{\partial H}{\partial u_1} = H'_1\lambda + \tfrac{1}{2}H_{1ij}u_i u_j + \cdots = 0 \tag{1.1.21}$$

and

$$\frac{\partial H}{\partial u_s} = H_{ss}u_s + H'_s\lambda + \tfrac{1}{2}H_{s11}u_1^2 + \cdots = 0 \tag{1.1.22}$$

where $s = 2, 3, \ldots, N$.

Solving (1.1.21) and (1.1.22) simultaneously and neglecting comparatively higher order terms result in the first order equations

$$H'_1\lambda + \tfrac{1}{2}H_{111}u_1^2 = 0$$

and
$$H_{ss}u_s + \left(H'_s - \frac{H'_1 H_{s11}}{H_{111}}\right)\lambda = 0 \tag{1.1.23}$$

which describe the first order approximation of the equilibrium path in the vicinity of the critical point F. On a plot of λ against the critical coordinate u_1, one observes a smooth extremum while $\lambda - u_s$ plot exhibits a straight line in this approximation and a sharp cusp if additional terms are considered (Figure 1.2). Recalling the fact that the equilibrium paths

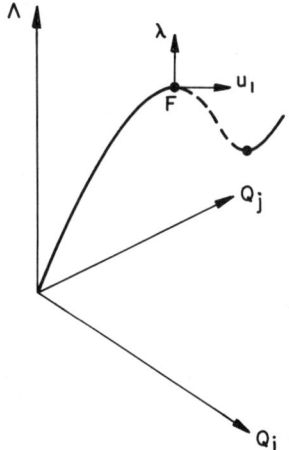

Figure 1.2

1.1 Simple (discrete) critical points

are space curves, one can envisage the $u_1 - \lambda$ and $u_s - \lambda$ plots as two distinct views of this curve from different angles. Clearly, the path has a limiting form in the vicinity of F, and such a critical point is referred to as a *limit point*.

Thus, if F is moved along the initially stable fundamental equilibrium path emerging from the origin of the load-deflection space, the stability coefficients H_{ii} which are positive in the beginning change *continuously* with the increasing Λ, and one of them passes through zero at a critical state. This is, then, a primary critical point at which H_{11} vanishes while all $H_{ss} > 0$ for $s \neq 1$, and it is clear from the above discussion that if all the other energy coefficients are finite (mathematically most general case) this critical point is a limit point. If F is moved further on the path, then H_{11} becomes negative and the path after the limit point is unstable. In Figure 1.2, a continuous line represents a stable path and a broken line an unstable path. It can readily be shown that due to nonvanishing cubic coefficient H_{111} the limit point itself is also unstable, and a structural system exhibits *snap-buckling* when such a point is reached.

The first equation of (1.1.23) indicates that the coefficient H'_1 plays an important role in the buckling behaviour of the system. Suppose now that this coefficient vanishes simultaneously with H_{11} while the remaining coefficients are all finite. The first order equilibrium equations, then, take the form

$$\tfrac{1}{2}H_{111}u_1^2 + cu_1\lambda + \tfrac{1}{2}d\lambda^2 = 0 \tag{1.1.24}$$

and

$$H_{ss}u_s + H'_s\lambda = 0 \tag{1.1.25}$$

where

$$c = H''_{11} - \frac{H_{s11}H'_s}{H_{ss}}$$

$$d = H''_1 - 2\frac{H'_{1s}H'_s}{H_{ss}} + \frac{H_{1sr}H'_s H'_r}{H_{ss}H_{rr}}$$

Equation (1.1.24) can be solved for u_1 to give

$$u_1 = \frac{1}{H_{111}}[-c \pm (c^2 - H_{111}d)^{\tfrac{1}{2}}]\lambda \tag{1.1.26}$$

which indicates that two distinct lines intersect each other at F provided

$$c^2 - H_{111}d > 0$$

Such a point is known as a *point of bifurcation* or *branching point*. On a plot of λ against u_1, the intersection appears distinctly while in the $u_s - \lambda$ plot the paths seem to be tangential (Figure 1.3). As in the case of the limit

1.1.2 Energy expansion and equilibrium paths

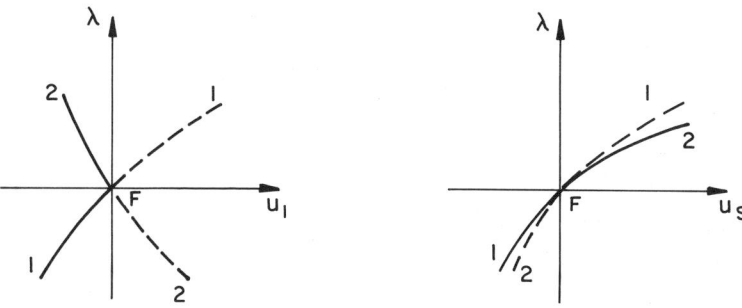

Figure 1.3

point, these plots represent two distinct views of the intersecting space curves from different angles.

Moving along the fundamental path, H_{11} becomes negative on passing through the critical point and the path loses its stability, while the other path gains stability on passing through F. In the terminology of Poincaré an exchange of stabilities occurs. Stability distribution on the equilibrium paths in the vicinity of a bifurcation point will be studied in Section 1.1.7.

The path intersecting the fundamental path is often referred to as the *post-buckling path* which generally has a finite slope, and the point of bifurcation is, therefore, called *asymmetric*. In special, but not uncommon situations, the post-buckling path will have a zero slope and the point F will, then, be called *stable symmetric* or *unstable symmetric* depending on whether the curvature is positive or negative respectively. Detailed discussion of these conditions will be deferred to Section 1.1.6.

Here, it is finally observed that a linear eigenvalue problem arises when all the coefficients H'_i, H''_i, etc., vanish at the unloaded state. The fundamental path, then, coincides with the Λ axis, and moving along the path a point of bifurcation is reached at which a second and distinct path is intersected. Supposing now that F is at such a critical point where $H_{11} = 0$ and $H'_i = H''_i = \cdots = 0$, the first order equilibrium equations take the form

$$u_1(\tfrac{1}{2}H_{111}u_1 + H'_{11}\lambda) = 0$$

in $u_1 - \lambda$ subspace, and

$$u_s = 0 \quad \text{for} \quad u_1 = 0$$

and

$$H_{ss}u_s + 2\frac{H'_{11}}{H_{111}}\left(\frac{H'_{11}}{H_{111}}H_{s11} - H'_{s1}\right)\lambda^2 = 0 \quad \text{for} \quad u_1 = -\frac{2H'_{11}}{H_{111}}\lambda$$

in $u_s - \lambda$ plane.

1.1 Simple (discrete) critical points

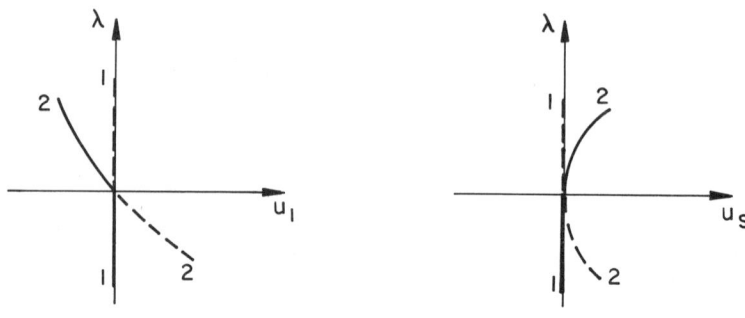

Figure 1.4

Figure 1.4 shows the fundamental and post-buckling paths in $u_1 - \lambda$ and $u_s - \lambda$ planes.

The loss of stability of many linear systems (such as Euler column) is associated with this type of pure eigenvalue problem, pre-buckling deflections being absent or negligible.

1.1.3 Perturbation methods

The preceding section serves two purposes; firstly, it helps to form a general opinion about the overall behaviour of the system in the load-deflection space, thus providing a basis for more, further detailed, studies. Secondly, it demonstrates the use of Taylor's expansions in exploring the path configurations in the vicinity of an equilibrium point. The stability of the paths and other related phenomena can also be studied by means of this method [36].

It is observed, however, that the derivation of the first order equations of an equilibrium path may require some skill with regard to retaining and/or neglecting a certain term compared to others. As the degree of nonlinearity increases and the effect of other factors, such as imperfections is to be accounted for, this procedure might pose certain algebraic difficulties. A more systematic method, therefore, seems desirable and as remarked earlier perturbation techniques suggest themselves as ideal tools for this type of analysis. As a matter of fact, perturbation theory in mechanics has been developed in an effort to simplify, in a mathematical way, a complicated problem. The behaviour of a nonlinear system can thus be studied conveniently in a sequence of ordered linear perturbation problems. An account of the present state of knowledge can be found in recent books by Van Dyke [37], Bellman [38] and Cole [39]. These books give a wide variety of formal perturbation procedures available in applied mathematics.

1.1.3 Perturbation methods

For the specific needs of the stability analyses here, Sewell [40] has outlined a 'static perturbation technique' in which equilibrium paths in the vicinity of a presumably known equilibrium state are sought in terms of a *single* perturbation parameter. This technique and its variations have been applied successfully by Sewell [42], Thompson [41] and others. In developing a general theory of stability concerned with multiple-parameter systems, however, it has been observed [43] that a one-parameter perturbation approach does not yield full information about the behaviour of such systems. Several new nonlinear concepts, such as the concept of an *equilibrium surface*, arise which can best be explored via a *multiple-parameter perturbation technique*. Such a method has recently been developed and employed in several applications [43–53]; the method reduces to the 'static perturbation technique' in the special case of one-parameter systems, and it will now be outlined.

Suppose $f(x_i)$ is a single-valued differentiable function of several variables x_i ($i = 1, 2, \ldots, N$), and there exist n ($n < N$) relations between these variables in the form

$$f_i(x_j) = 0 \quad i = 1, 2, \ldots, n; \quad j = 1, 2, \ldots, N \tag{1.1.27}$$

Suppose further that the above n relations can be solved simultaneously to yield solutions in the parametric form

$$x_i = x_i(\eta^\alpha) \tag{1.1.28}$$

where η^α ($\alpha = 1, 2, \ldots, N - n$) are certain real independent parameters which are chosen such that the functions (1.1.28) are single-valued. From a geometrical point of view (1.1.28) may be considered as describing some $(N - n)$ dimensional surfaces in the N-dimensional x_i-space. One can explore the neighbourhood of a presumably known point on one of these surfaces in a sequence of ordered perturbation problems.

Thus, one seeks to construct the functions (1.1.28) term by term sequentially in the form

$$x_i - x_i^\circ = x_i^\alpha \eta^\alpha + \frac{1}{2!} x_i^{\alpha\beta} \eta^\alpha \eta^\beta + \cdots \tag{1.1.29}$$

where $\eta^\alpha = 0$ defines the known point x_i°, and superscripts on the x_i denote partial differentiation with respect to the corresponding η^α, all derivatives x_i^α, $x_i^{\alpha\beta}$ etc. being evaluated at the point x_i°. Here and in the sequel the summation convention applies to both superscripts and subscripts separately, unless otherwise stated.

It is assumed here that the surface(s) of interest in the vicinity of x_i° can be expressed in the series form (1.1.29). To obtain the coefficients of (1.1.29), substitute first the assumed solutions (1.1.28) back into the

1.1 Simple (discrete) critical points

defining equations $f_i = 0$ to have the identities

$$f_i[x_i(\eta^\alpha)] \equiv 0, \tag{1.1.30}$$

and then differentiate these identities with respect to the η^α as many times as may be required. It is important to note that the f_i are now identically zero functions of the independent parameters η^α and successive differentiations of a particular $f_i \equiv 0$ with respect to any η^α will always yield a differentiated form equal to zero.

Thus, differentiating (1.1.30) with respect to η^α ($\alpha = 1, 2, \ldots, N-n$) and evaluating at x_i^0 yield

$$f_{ij}x_j^\alpha = 0 \tag{1.1.31}$$

A second differentiation with respect to η^β ($\beta = 1, 2, \ldots, N-n$) yields on evaluation

$$f_{ijk}x_j^\alpha x_k^\beta + f_{ij}x_j^{\alpha\beta} = 0 \tag{1.1.32}$$

Similarly, third, fourth and higher order equations can be obtained by successive differentiations and evaluations.

Allowing now $(N-n)$ *appropriate* basic variables, x_α say, to take on the role of the η^α, one observes that at each step of the sequence the number of the equations is equal to the number of unknowns since $\partial x_\alpha/\partial x_\alpha = 1$, $\partial x_\alpha/\partial x_\beta = 0$, $\partial^2 x_\alpha/\partial x_\alpha \partial x_\beta = 0$ due to the assumption that the x_α are now independent variables.

Thus, the first perturbation produces $n(N-n)$ linear equations (1.1.31) which are solved simultaneously for the same number of unknowns, the first derivatives $x_i^\alpha \equiv \partial x_i/\partial x_\alpha|_{x_i = x_i^0}$. These first derivatives are then substituted into the second order equations (1.1.32) which now represent a set of linear equations in the second derivatives $x_i^{\alpha\beta}$ and are, therefore, solved readily.

The procedure follows similarly, each step supplying the coefficients of (1.1.29) successively. Thus, the surface around the known point can be explored to the desired extent.

In the choice of the independent perturbation parameters, some care must be exercised. The general guideline is that the surface in the vicinity of the known point must be represented by single-valued functions of these parameters so that a continuous progress across the surface is described when the parameters take on increasing values. A basic variable associated with an axis which is tangential to the surface normally represents a good choice, and in this regard appropriate coordinate transformations may simplify the analysis (see, for example, Section 2.2.2).

Here, the formal steps of the technique have been outlined in general terms; details of the specific applications to various aspects of the stability theory will be given through the sections. Starting with the following section, the analyses in Part 1 will be based mostly on one-parameter perturbations while in Part 2 and certain sections of Part 1, multiple-parameter techniques will be employed.

1.1.4 Stability of an equilibrium state

Returning now to the basic question concerning the stability of an equilibrium state, consider an arbitrary point of equilibrium, F, at which the potential energy has the form (1.1.10).

Using the suggestive language of geometry, it may be asserted that the energy surface around F in the u_i-space must be a locally *strictly convex* surface whose concavity is upwards if F is to represent a minimum. The convexity of the surface can be examined through the variations of the energy function (1.1.10) with respect to an arbitrary path in the u_i-space of the form

$$u_i = u_i(\eta) = u_{i,\eta} + \tfrac{1}{2} u_{i,\eta\eta} \eta^2 + \cdots \quad (1.1.33)$$

where η is an unspecified path parameter, $\eta = 0$ defines the equilibrium point F, and subscripts η on u's denote differentiation with respect to η [43, 52, 53].

To examine the variations of the energy function $H(u_i, \lambda)$ along this path, the change in energy is expressed as

$$h(\eta) = H[u_i(\eta), 0] - H_0(0, 0) \quad (1.1.34)$$

Then,

$$\frac{\mathrm{d}h(\eta)}{\mathrm{d}\eta} = H_i u_{i,\eta}$$

which upon evaluation at F yields

$$\left. \frac{\mathrm{d}h}{\mathrm{d}\eta} \right|_F = 0$$

since $H_i = 0$ at the equilibrium state F.

Differentiating (1.1.34) twice yields

$$\frac{\mathrm{d}^2 h}{\mathrm{d}\eta^2} = H_{ij} u_{i,\eta} u_{j,\eta} + H_i u_{i,\eta\eta}$$

which upon evaluation takes the form

$$\left. \frac{\mathrm{d}^2 h}{\mathrm{d}\eta^2} \right|_F = H_{ii}(u_{i,\eta})^2 \quad (1.1.35)$$

1.1 Simple (discrete) critical points

Evidently, if all $H_{ii} > 0$, the second variation

$$\delta^2 h = \frac{1}{2} \frac{d^2 h}{d\eta^2}\bigg|_F \eta^2$$

remains positive for all possible paths (1.1.33), thus ensuring a strictly convex surface whose concavity is upwards and indicating that F represents a minimum. If one or more H_{ii} is negative, the arbitrary path (1.1.33), which is yet in the form of a straight line, can be oriented in the corresponding direction(s) to yield a negative second variation(s), thus ruling out the possibility of a minimum and hence stability.

Attention will now be focussed on the indecisive case in which F is primary critical.

Assume first that the critical point is simple (discrete) and that $H_{11} = 0$ while $H_{ss} > 0$ for $s = 2, 3, \ldots, N$. Writing (1.1.35) in the form

$$\frac{d^2 h}{d\eta^2}\bigg|_F = H_{11}(u_{1,\eta})^2 + H_{ss}(u_{s,\eta})^2$$

it is observed that a ray in the u_1 direction ($u_{s,\eta} = 0$) yields a zero second variation while the rays in all other directions yield a positive second variation. To establish the convexity of the surface, therefore, it is necessary to determine the higher order variations with respect to a path emerging initially as a ray in the u_1 direction. The role of η can now be given to u_1 for convenience [43, 52, 53], and the path (1.1.33), then, has the first path derivatives $u_{1,1} = 1$ and $u_{s,1} = 0$ ($s = 2, 3, \ldots, N$).

The third differentiation of (1.1.34) with respect to u_1 ($\equiv \eta$) gives

$$\frac{d^3 h}{du_1^3} = H_{ijk} u_{i,1} u_{j,1} u_{k,1} + 3 H_{ij} u_{i,1} u_{j,11} + H_i u_{i,111}$$

which yields

$$\frac{d^3 h}{du_1^3}\bigg|_F = H_{111}$$

A *necessary condition* for stability is that this cubic coefficient vanishes. If it does not vanish the analysis need not be pursued any further, the possibility of a strictly convex surface being ruled out.

Mathematically it might seem improbable that H_{111} should vanish, but actually, due to certain symmetry properties, this coefficient often vanishes and the determination of the fourth variation is required for a decision regarding stability.

Differentiate (1.1.34) for a fourth time, then, to get

$$\frac{d^4 h}{du_1^4} = H_{ijkl} u_{i,1} u_{j,1} u_{k,1} u_{l,1} + 6 H_{ijk} u_{i,1} u_{j,1} u_{k,1}$$

$$+ 3 H_{ij} u_{i,11} u_{j,11} + 4 H_{ij} u_{i,1} u_{j,111} + H_i u_{i,1111}$$

1.1.4 Stability of an equilibrium state

which, under the condition $H_{111} = 0$, yields

$$\left.\frac{d^4h}{du_1^4}\right|_F = H_{1111} + 6H_{s11}u_{s,11} + 3H_{ss}(u_{s,11})^2 \qquad (1.1.36)$$

It is immediately noted that the second path derivatives $u_{s,11}$ of the path (1.1.33) are now in the picture. The first derivatives $u_{1,1} = 1$ and $u_{s,1} = 0$ were predicted in the course of analysis, and were dominant quantities in the second and third variations, dispensing with the second path derivatives, i.e., a *ray* was yielding sufficient information so far, but for the evaluation of the fourth variation of the energy function the second path derivatives $u_{s,11}$ are needed.

The particular path(s) along which the energy surface remains closest to the u_i-space are sought since such a path has the potential to reveal a negative variation (if any) of the energy function. At this stage, the path(s) associated with a minimum variation of energy can be identified by minimizing (1.1.36) with respect to $u_{s,11}$. Thus, differentiating (1.1.36) with respect to $u_{r,11}$ ($r = 2, 3, \ldots, N$) and equating to zero yield

$$6H_{r11} + 6H_{rr}u_{r,11} = 0 \qquad \text{(no summation over } r\text{)}$$

which results in

$$u_{r,11} = -\frac{H_{r11}}{H_{rr}} \qquad (r = 2, 3, \ldots, N) \qquad (1.1.37)$$

Substituting for $u_{r,11}$ into (1.1.36) from (1.1.37) gives

$$\left.\frac{d^4h}{du_1^4}\right|_F = H_{1111} - 3\sum_{s=2}^{N}\frac{(H_{s11})^2}{H_{ss}} \qquad (1.1.38)$$

which leads to the fourth variation

$$\delta^4 h = \frac{1}{4!}\left.\frac{d^4h}{du_1^4}\right|_F u_1^4$$

If the derivative (1.1.38) is positive the critical state F is stable and if this derivative is negative, F is unstable.

Although unlikely, it might happen that the fourth derivative (1.1.38) vanishes, rendering the situation indecisive once more. Then, the vanishing of the fifth variation is a necessary condition for stability, and if satisfied, the positive definiteness of the sixth variation becomes a sufficient condition. The procedure of examining higher variations of $h(u_1)$ is now clear and the analysis will not be pursued any further.

It is worth noting that the path derivatives (1.1.37) can also be determined at the onset of the analysis independently. To do so, one recalls that among the paths given in the form (1.1.33), the one along

1.1 Simple (discrete) critical points

which the energy surface remains in contact with the u_i-space to the highest order is of the utmost importance. In fact, the notion of 'order of contact' has been used by Sewell to discuss the stability of an equilibrium point, and to obtain such paths. Thus, observing that the necessary conditions for a minimum are given by $H_i = 0$, the assumed path $u_i = u_i(u_1)$ is substituted into these conditions to get the identities

$$H_i[u_j(u_1), 0] \equiv 0$$

Differentiating these identities with respect to u_1 yields

$$H_{ij}u_{j,1} = 0$$

which upon evaluation at F results in

$$u_{s,1} = 0 \quad \text{for} \quad s = 2, 3, \ldots, N;$$

the first derivatives predicted earlier in the course of the analysis.

The second differentiation yields

$$H_{ijk}u_{j,1}u_{k,1} + H_{ij}u_{j,11} = 0$$

which results in

$$H_{111} = 0 \quad \text{for} \quad i = 1$$

and

$$u_{s,11} = -\frac{H_{s11}}{H_{ss}} \quad \text{for} \quad s = 2, 3, \ldots, N;$$

the second derivatives (1.1.37) which are meaningful only when H_{111} vanishes.

The advantage of this procedure is that one can evaluate the path derivatives and construct the path in advance. The analysis, then, consists of sequential perturbations of (1.1.34) and $H_i[u_j(u_1), 0] \equiv 0$ as may be required.

The stability of coincident critical points can be examined via the multiple-parameter perturbation technique similarly, and will be presented in Section 1.5.1.

The post-bifurcation behaviour of a system depends on the stability or instability of the critical bifurcation point itself and the results of the present section will be discussed in the context of post-critical behaviour at the end of Section 1.1.6.

1.1.5 Critical loads

It is clear from the discussion in Section 1.1.2 that a stable fundamental equilibrium path passing through the origin of the load-deflection space

1.1.5 Critical loads

normally loses its stability at a bifurcation or limit point. The form of equilibrium paths and their stability in the vicinity of a critical point will be studied in greater detail via the systematic perturbation technique. Before embarking on such an essentially post-buckling analysis, however, it seems desirable to discuss first the critical points themselves and explore the methods of determining or estimating the critical loads. Such a buckling analysis has, in fact, been considered as the sole problem of elastic stability for two centuries until increasing demands for economy necessitated a thorough understanding of post-buckling behaviour.

In determining the elastic buckling loads of practical structures, a traditional approach is to neglect the deformations that occur prior to buckling. In the load-deflection space, this situation is represented by a fundamental path which coincides with the loading axis, generalized coordinates remaining zero until the loading parameter reaches a critical value at which buckling deformations develop. This approximation can be justified in predicting critical branching loads of many structural systems such as axially loaded columns, frames, plates, etc.; it is clear, however, that limit points can not be located by such an approach. Besides, it leads to severe errors in certain important problems such as the buckling of a shallow arch and shallow spherical shell in which *pre-buckling* deflections play a significant role in the branching behaviour of the system. In the load-deflection space, one is faced with the problem of locating a point of bifurcation on a nonlinear fundamental path.

A method of branching analysis, taking into account the effect of pre-buckling deflections has been proposed by Masur and Schreyer [55] who developed the basic equations of their theory in the context of nonlinear elasticity. In this analysis, a second approximation of the critical bifurcation load is obtained conveniently by using the Lagrange multipliers. Thompson [54] has outlined a perturbation approach which yields an ordered form of stability estimates for discrete systems. For symmetric systems, a more convenient perturbation procedure was described in Refs. [43] and [50].

In this section a general perturbation technique designed to yield ordered estimates for limit points as well as bifurcation points is discussed. Symmetric systems will be treated in Sections 1.3.3 and 1.3.4.

Consider the system characterized by the potential energy function (1.1.1). Suppose the equilibrium equations $V_i = 0$ have been solved simultaneously to yield the fundamental equilibrium path through the origin of load-deflection space in the parametric form

$$Q_i = Q_i(\eta), \quad \Lambda = \Lambda(\eta)$$

where η is chosen such that these functions are single-valued.

1.1 Simple (discrete) critical points

Substituting this assumed solution back into the equilibrium equations results in the identities

$$V_i[Q_i(\eta), \Lambda(\eta)] \equiv 0 \qquad (1.1.39)$$

Differentiating (1.1.39) with respect to η and evaluating at the origin $Q_i = \Lambda = \eta = 0$, one has

$$V_{ij}Q_{j,\eta} + V'_i\Lambda_\eta = 0 \qquad (1.1.40)$$

A second differentiation and evaluation yields

$$(V_{ijk}Q_{k,\eta} + V'_{ij}\Lambda_\eta)Q_{j,\eta} + V_{ij}Q_{j,\eta\eta}$$
$$+ (V'_{ij}Q_{j,\eta} + V''_i\Lambda_\eta)\Lambda_\eta + V'_i\Lambda_{\eta\eta} = 0$$

The third, fourth, etc. order equations can similarly be obtained by successive perturbations of (1.1.39).

Allowing now one of the basic variables to take on the role of η, one observes that (1.1.40) represents a set of N linear equations in N first derivatives. If, for example, Λ is equated to η, (1.1.40) takes the form

$$V_{ij}Q'_j + V'_i = 0$$

which can be solved for Q'_j, a prime on the Q_j reflecting the fact that η is now replaced by Λ. Also note that in structural problems V_{ij} is positive definite since the origin represents a stable equilibrium state, and the path through the origin is, therefore, *unique*. It is further noted in passing that this last statement will evidently hold true for any stable equilibrium state and not just for the origin, as shown earlier also with the aid of the energy expansion.

Having obtained the first derivatives, one then substitutes them into the second order equations which after setting $\eta = \Lambda$ take the form

$$(V_{ijk}Q'_k + V'_{ij})Q'_j + V_{ij}Q''_j + V'_{ij}Q'_j + V''_i = 0 \qquad (1.1.41)$$

It is seen that (1.1.41) now represents a linear set of equations in the second derivatives Q''_j which can readily be solved. Due to the fact that the Q''_j are associated again with the positive definite matrix V_{ij}, the uniqueness of solution is carried over to second derivatives, and similarly to higher order derivatives successively.

It must be pointed out that the loading parameter Λ is not always an appropriate perturbation parameter, and often one of the generalized coordinates may be required to take on this role although the choice of Λ simplifies the analysis considerably. In the case of limit points, for instance, Λ becomes an inappropriate choice since the path can not be expressed as a single-valued function of Λ. Besides, it will be seen in the

1.1.5 Critical loads

analysis of symmetric systems (Chapter 1.3) that even for bifurcation points, the choice of an appropriate generalized coordinate as the perturbation parameter improves the chances of obtaining a faster and better estimate of the critical load in the ordered scheme of approximations. This is partly due to the fact that in many nonlinear structural problems the fundamental path is likely to take the typical form illustrated in Figure 1.1 which can be expressed in terms of a suitable generalized coordinate more adequately and more fully. Nevertheless, if the fundamental path is moderately nonlinear the choice of Λ as the perturbation parameter yields a simpler symmetric treatment.

Having evaluated the path derivatives sequentially, the fundamental path can be constructed in the series form to the degree of approximation desired. Thus, using the general parametric form again one has

$$Q_i = Q_i(\eta) = Q_{i,\eta}\eta + \frac{1}{2!}Q_{i,\eta\eta}\eta^2 + \cdots$$
$$\Lambda = \Lambda(\eta) = \Lambda_\eta \eta + \frac{1}{2!}\Lambda_{\eta\eta}\eta^2 + \cdots$$
(1.1.42)

The fundamental path described by (1.1.42) is initially stable, and any loss of stability on the path can be identified by the vanishing of the stability determinant,

$$\det |V_{ij}| = 0 \tag{1.1.43}$$

The points on the path which satisfy this determinant can be limit points or bifurcation points, and the following general analysis is designed to locate both types of critical points in a scheme of ordered approximations. Attention, however, will be directed specifically to bifurcation points for which the loading parameter may take the role of η. The reason for this lies in the fact that limit points can also be obtained from the preceding equilibrium analysis since they represent the extremum points on the estimated paths.

The variation of the stability determinant

$$\Delta(Q_k, \Lambda) \equiv \det |V_{ij}(Q_k, \Lambda)| \tag{1.1.44}$$

along the fundamental path (1.1.42) will now be examined. The progress along the path is, in general, described by η, and the variation of the determinant (1.1.44) can likewise be studied by means of this parameter. Thus the determinant along the path is expressed as

$$A(\eta) \equiv \Delta[Q_i(\eta), \Lambda(\eta)] \tag{1.1.45}$$

which can be differentiated and evaluated at the origin successively to

1.1 Simple (discrete) critical points

obtain the ordered derivatives of A. These derivatives are

$$A_\eta = \Delta_i Q_{i,\eta} + \Delta' \Lambda_\eta$$
$$A_{\eta\eta} = (\Delta_{ij} Q_{j,\eta} + \Delta'_i \Lambda_\eta) Q_{i,\eta} + \Delta_i Q_{i,\eta\eta}$$
$$+ (\Delta'_i Q_{i,\eta} + \Delta'' \Lambda_\eta) \Lambda_\eta + \Delta' \Lambda_{\eta\eta} \qquad (1.1.46)$$
$$A_{\eta\eta\eta} = \cdots$$

One of the advantages of this approach is that the path derivatives $Q_{i,\eta}$, $Q_{i,\eta\eta}$, etc. determined in the equilibrium analysis are now readily available for determining the above stability derivatives.

If Λ is given the role of η, then, (1.1.46) reads

$$A' = \Delta_i Q'_i + \Delta' \qquad (1.1.47)$$
$$A'' = (\Delta_{ij} Q'_j + \Delta'_i) Q'_i + \Delta_i Q''_i + \Delta'_i Q'_i + \Delta''$$

etc.

The stability determinant referred to the origin can now be expressed in the general parametric form

$$A(\eta) = A_0 + A_\eta \eta + \frac{1}{2!} A_{\eta\eta} \eta^2 + \cdots$$

where

$$A_0 = A(0) = \det |V_{ij}(0, 0)|.$$

Truncations of this series after the second, third, etc. terms yield the first, second, etc. order stability equations as

$$A_0 + A_\eta \eta = 0$$
$$A_0 + A_\eta \eta + \frac{1}{2!} A_{\eta\eta} \eta^2 = 0 \qquad (1.1.48)$$

etc.

which can be solved with the equilibrium equations concurrently to obtain the first, second, etc. order estimates for the critical load.

It is noted again that if Λ is given the role of η, (1.1.48) takes the form

$$A_0 + A' \Lambda = 0$$
$$A_0 + A' \Lambda + \frac{1}{2!} A'' \Lambda^2 = 0$$

etc.

and the estimates of the critical load are obtained directly.

In discussing the buckling behaviour of symmetric systems in

1.1.6 Post-critical behaviour

Section 1.3.4, one of the generalized coordinates will be employed as the independent perturbation parameter and more explicit results will be obtained.

1.1.6 Post-critical behaviour

In this section, exploration of equilibrium paths, their shape and stability is continued, attention being focussed on the behaviour of the system in the vicinity of a bifurcation point.

It was demonstrated in Section 1.1.2 that limit points are mathematically more general and arise when a single stability coefficient vanishes while bifurcation points arise when a stability coefficient and a second energy coefficient vanish simultaneously [35]. In specific problems, limit points are common when the system has *imperfections*, and it is well known that sufficiently shallow arches and spherical caps represent two typical structures which may exhibit limit points in their *perfect* forms. Perfect systems, however, are often associated with bifurcation points which manifest themselves mainly in three forms: *asymmetric, stable symmetric* and *unstable symmetric* (Figures 1.5a,b,c). It was seen in Section 1.1.2 that asymmetric points of bifurcation are again mathematically more general than the other two since they arise whenever a stability coefficient vanishes together with a related energy coefficient and the

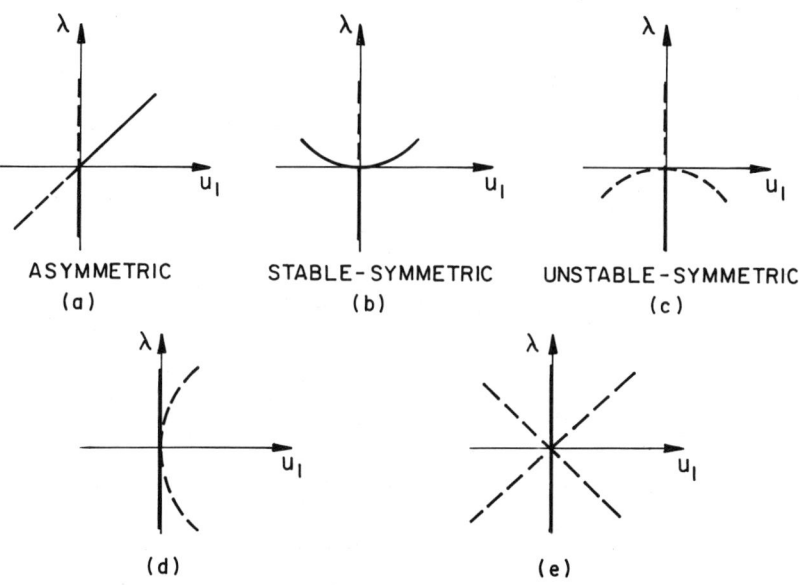

Figure 1.5

1.1 Simple (discrete) critical points

slope of the associated post-buckling is in general finite. Rigidly-jointed triangular frames [27, 36], oblate spheroidal shells under external pressure [56], for example, exhibit asymmetric points of bifurcation.

Symmetric points of bifurcation which are associated with an initially flat post-buckling path are more common in engineering problems despite the fact that analytically an additional coefficient must vanish simultaneously. It seems that structural systems are provided with sufficient symmetry often resulting in a behaviour which seems improbable mathematically. Examples are Euler column (stable-symmetric), plates loaded in their plane (stable-symmetric), axially compressed cylindrical panels (both stable and unstable-symmetric), various forms of cylindrical shells under various loads (both stable and unstable-symmetric), shallow arch (unstable-symmetric), etc.

Essential features of these three commonly arising bifurcation points including the effect of small imperfections were first examined by Koiter in his thesis. Roorda [36, 57–59], in presenting his theory concerning various types of imperfections, has treated the subject in terms of generalized coordinates and by using potential energy expansions coupled with appropriate transformations. Thompson and Sewell discussed the behaviour of the system in the vicinity of the three branching points by means of perturbation techniques.

Recently, the multiple-parameter theory [43] revealed that, in addition to the commonly arising bifurcation points above, two more interesting phenomena may arise under some circumstances (Figures 1.5d,e). The situation illustrated in Figure 1.5d takes place when *one additional* coefficient vanishes while the paths illustrated in Figure 1.5e arise under certain symmetry conditions associated with *two additional* vanishing coefficients. Both cases were first discussed fully in the context of multiple-parameter systems in [43], and later in [48] and [49]. If the system has more than one independent loading parameter, the conditions giving rise to these phenomena can be induced by suitably combining these parameters. In fact, it has been shown analytically in [43], [48] and [49], that such critical points may occur if the loading follows a particular path which is tangential to the *stability boundary* (defined in Section 2.1.2). In one-parameter systems, however, such situations may rarely occur, and an analytical treatment will be given in the context of multiple-parameter systems (see Chapters 2.3 and 2.4).

Thus, this section is devoted to the analysis of the three commonly arising bifurcation points.

Consider the structural system described by

$$V = V(Q_i, \Lambda)$$

1.1.6 Post-critical behaviour

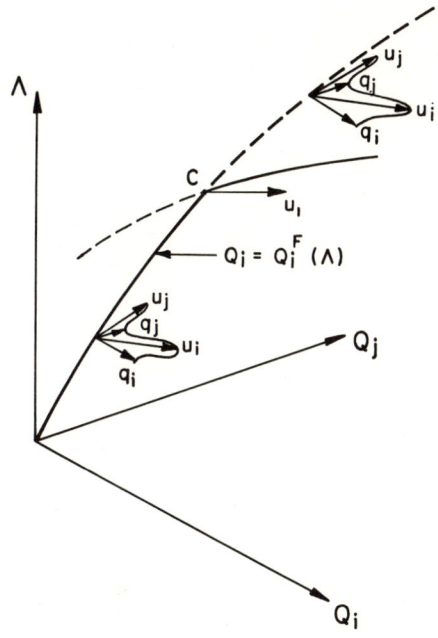

Figure 1.6

as in the preceding sections, and suppose that the equilibrium equations have been solved simultaneously to yield the fundamental path $Q_i = Q_i^F(\Lambda)$. Since attention here will be focussed on the neighbourhood of a bifurcation point on the fundamental path, it may be assumed that $Q_i = Q_i^F(\Lambda)$ is single-valued in the region of interest, this being an essential requirement of the following analysis (Figure 1.6). The potential energy can, then, be referred to the fundamental path by setting

$$Q_i = Q_i^F(\Lambda) + q_i \qquad (1.1.49)$$

which indicates that the small incremental coordinates q_i are now measured from the fundamental path (Figure 1.6).

A further change of coordinates by means of a linear orthogonal transformation,

$$q_i = \alpha_{ij}(\Lambda)u_j, \qquad \alpha_{ij}(\Lambda)\alpha_{jk}(\Lambda) = \delta_{ik}, \qquad (1.1.50)$$

is introduced to diagonalize the quadratic form in the q_i.

Introducing (1.1.49) and (1.1.50) into the energy function one obtains the transformed function

$$S(u_i, \Lambda) \equiv V[Q_i^F(\Lambda) + \alpha_{ij}u_j, \Lambda] \qquad (1.1.51)$$

1.1 Simple (discrete) critical points

with the properties

$$S_i(0, \Lambda) = S'_i(0, \Lambda) = S''_i(0, \Lambda) = \cdots = 0 \tag{1.1.52}$$

and

$$S_{ij}(0, \Lambda) = S'_{ij}(0, \Lambda) = S''_{ij}(0, \Lambda) = \cdots = 0 \quad \text{for} \quad i \neq j \tag{1.1.53}$$

where subscripts and a prime on S denote partial differentiation with respect to the u_i and Λ respectively. The first of these properties follows immediately from the fact that $u_i = 0$ defines the fundamental path, and the second is due to the transformation (1.1.50) which is a function of Λ so that the quadratic form is diagonalized at every point on the path. In other words, a given value of Λ defines an equilibrium state on the fundamental path and a transformation matrix $\alpha_{ij}(\Lambda)$ which diagonalizes the quadratic form of the energy at that point. As the parameter Λ changes, the u_i axes slide along the fundamental path and rotate accordingly so that the quadratic form is kept diagonalized at every point on the path (Figure 1.6). Note that by insisting that the fundamental path is single-valued in the region of interest, and effectively by introducing the transformation (1.1.49) which is essentially different from that of Section 1.1.2, limit points have automatically been left out of the scope of the analysis.

Thus, to explore the equilibrium paths in the vicinity of a simple bifurcation point C lying on the fundamental path, suppose that such a point is given by $S_{11}(0, \Lambda^c) = 0$ and $S_{ss}(0, \Lambda^c) \neq 0$ for $s = 2, 3, \ldots, N$ where Λ^c is the critical value of Λ. Following the perturbation procedure outlined in Section 1.1.3, the solution of the equilibrium equations $S_i(u_j, \Lambda) = 0$ are formally expressed in parametric form $u_i = u_i(\eta)$, and $\Lambda = \Lambda(\eta)$. However, under the assumptions of this section and from physical considerations the critical coordinate u_1 is an appropriate perturbation parameter.

Thus, assuming that any post-buckling path emanating from the critical point C is in the form

$$u_s = u_s(u_1) \quad \text{and} \quad \Lambda = \Lambda(u_1) \tag{1.1.54}$$

and substituting back into the equilibrium equations $S_i = 0$ one obtains the identities

$$S_i[u_s(u_1), \Lambda(u_1), u_1] \equiv 0 \tag{1.1.55}$$

Differentiating (1.1.55) with respect to u_1 yields

$$S_{is}u_{s,1} + S'_i\Lambda_1 + S_{i1} = 0.$$

Upon evaluation at the critical point C where $u_i = 0$ and $\Lambda = \Lambda^c$, it is seen that for $i = 1$ the equation is identically satisfied while for $i = s$

1.1.6 Post-critical behaviour

$(s = 2, 3, \ldots, N)$ one obtains the first derivatives

$$u_{s,1} = 0 \qquad (1.1.56)$$

Differentiating (1.1.55) with respect to u_1 for a second time gives

$$(S_{isr}u_{r,1} + S'_{is}\Lambda_1 + S_{is1})u_{s,1} + S_{is}u_{s,11}$$
$$+ (S'_{is}u_{s,1} + S''_i\Lambda_1 + S'_{i1})\Lambda_1 + S'_i\Lambda_{11}$$
$$+ S_{i1s}u_{s,1} + S'_{i1}\Lambda_1 + S_{i11} = 0$$

which on evaluation at C yields

$$\Lambda_1 = -\frac{S_{111}}{2S'_{11}} \quad \text{for} \quad i = 1$$

and $\qquad (1.1.57)$

$$u_{s,11} = -\frac{S_{s11}}{S_{ss}} \quad \text{for} \quad i = s \neq 1$$

Using these derivatives and Taylor's expansion, the first order equations of the post-buckling path can be constructed as

$$S'_{11}\lambda + \tfrac{1}{2}S_{111}u_1 = 0$$

and $\qquad (1.1.58)$

$$S_{ss}u_s + \tfrac{1}{2}S_{s11}u_1^2 = 0$$

where

$$\lambda = \Lambda - \Lambda^c.$$

Evidently, the post-buckling path has a finite slope, and (1.1.58) is associated with an asymmetric point of bifurcation as discussed in Section 1.1.2. The only difference of course is that here the post-buckling path is referred to the fundamental path and consequently equations (1.1.58) are in a simpler form.

It is also understood that all relevant coefficients are assumed to be nonzero; the case in which $S'_{11} = 0$ is associated with the situation illustrated in Figure 1.5d. If S'_{11} and S_{111} vanish simultaneously and the energy function is symmetric in u_1, then, the situation shown in Figure 1.5e arises. As remarked earlier, however, these two exceptional cases will be discussed in Part 2.

The case in which only $S_{111} = 0$, on the other hand, is quite common and it frequently occurs due to certain symmetry properties of the system, thus resulting in symmetric points of bifurcation with $\Lambda_1 = 0$.

Proceeding to determine higher order derivatives in this special case, differentiate (1.1.55) with respect to u_1 for a third time and evaluate at C

1.1 Simple (discrete) critical points

to get

$$\Lambda_{11} = -\frac{1}{3S'_{11}}\left[S_{1111} - 3\sum_{s=2}^{N}\frac{(S_{s11})^2}{S_{ss}}\right], \quad (1.1.59)$$

the curvature of the post-buckling path. Depending on its sign, the bifurcation point C is either stable symmetric ($\Lambda_{11} > 0$) or *unstable symmetric* ($\Lambda_{11} < 0$). The asymptotic equation of the post-buckling path in $u_1 - \lambda$ plane is, then, in the form

$$S'_{11}\lambda + \frac{1}{3!}\left[S_{1111} - 3\sum_{s=2}^{N}\frac{(S_{s11})^2}{S_{ss}}\right]u_1^2 = 0 \quad (1.1.60)$$

which is plotted in Figure 1.7, together with the asymmetric point. The $u_s - u_1$ relationship obviously remains the same as in (1.1.58) and it is interesting to note that in the $u_s - u_1$ plane the path is tangential to u_1 indicating the passive role of the noncritical coordinates in the initial post-buckling behaviour (Figure 1.8). On a plot of λ against u_s or corresponding deflection, fundamental and post-buckling paths appear tangential as illustrated in Figure 1.3.

As shown in Figure 1.7, an exchange of stabilities between the fundamental and post-buckling paths takes place in the asymmetric case while the post-buckling paths are either totally stable or unstable in the symmetric cases. These phenomena will be shown analytically in the following section.

Problem. Using the function $H(u_i, \lambda)$ introduced in Section 1.1.2, perform a perturbation analysis to obtain the asymptotic equations of the

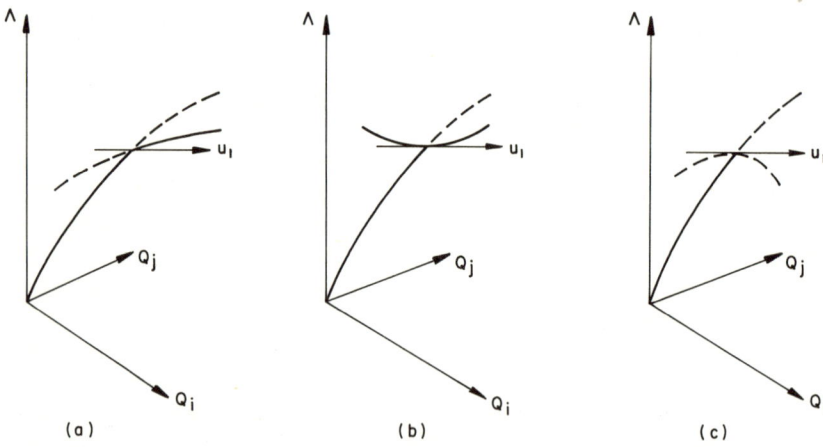

Figure 1.7

1.1.7 Stability distribution on the equilibrium paths

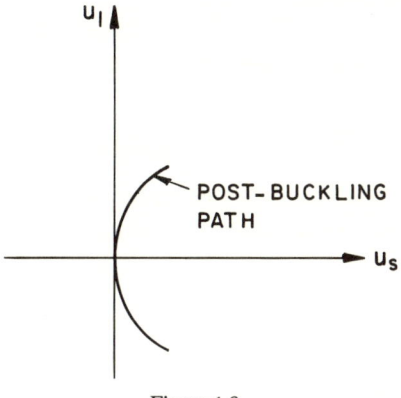

Figure 1.8

equilibrium path in the vicinity of a limit point, and rederive the related results of that section. It is understood that the function $S(u_i, \lambda)$ given by (1.1.51) cannot be used for this purpose. (Hint: Use u_1 as the perturbation parameter.)

1.1.7 Stability distribution on the equilibrium paths in the vicinity of a bifurcation point

The perturbation method was first adapted in [43] to study the stability distribution on the equilibrium paths in the vicinity of a *primary* critical point by examining the sign of the stability determinant

$$\Delta(u_i, \Lambda) \equiv |S_{ij}(u_i, \Lambda)| \qquad (1.1.61)$$

If an equilibrium state is noncritical a necessary and sufficient condition of stability is the positive definiteness of the second variation of energy. This implies that the determinant (1.1.61) evaluated at that point is positive. If the stability determinant is negative, then, the equilibrium state is unstable provided the state under consideration is in the neighbourhood of a *primary* critical point where $S_{11} = 0$ and $S_{ss} > 0$ ($s \neq 1$). It must be emphasized, however, that the assessment of stability by examining the sign of the stability determinant is only valid in the vicinity of a *primary* critical point since the stability coefficients cannot change sign in this small region before passing through zero or another critical point [43, 49].

The derivatives of the determinant (1.1.61) must first be evaluated. Thus, differentiating the determinant (1.1.61) by rows (or by columns) once with respect to u_i and once with respect to Λ and evaluating at the

1.1 Simple (discrete) critical points

critical point C one obtains

$$\Delta_i = S_{11i} \prod_{s=2}^{N} S_{ss} \quad \text{and} \quad \Delta' = S'_{11} \prod_{s=2}^{N} S_{ss} \tag{1.1.62a}$$

respectively. $\Delta_1 = S_{111} \prod_{s=2}^{N} S_{ss}$ will obviously vanish for symmetric points of bifurcation and the second derivative can similarly be obtained. Differentiating (1.1.61) twice and evaluating at the critical point yield

$$\Delta_{11} = \left[S_{1111} - 2 \sum_{s=2}^{n} \frac{(S_{s11})^2}{S_{ss}} \right] \prod_{s=2}^{N} S_{ss} \tag{1.1.62b}$$

For asymmetric points of bifurcation, the stability determinant can now be expanded into Taylor series around C as

$$\Delta = (S_{11i} u_i + S'_{11} \lambda) \prod_{s=2}^{N} S_{ss} + (\cdots) + \cdots \tag{1.1.63}$$

Evaluation of this determinant at an arbitrary point on the fundamental path (i.e., setting $u_i = 0$) yields to a first approximation

$$\Delta = S'_{11} \lambda \prod_{s=2}^{N} S_{ss} \tag{1.1.64}$$

If, $S'_{11} < 0$ say, then $\lambda < 0$ ($\lambda > 0$) defines stable (unstable) states while $\lambda = 0$ gives the critical point C. This is, of course, consistant with earlier remarks that a stable path loses its stability at a critical point where one of the positive stability coefficients becomes zero and subsequently changes its sign with a further increase in the loading parameter.

In order to examine the stability of states lying on the post-buckling path, the determinant (1.1.63) is evaluated at an arbitrary point on this path which is given by (1.1.58). Substituting for u_s and λ one obtains to a first approximation

$$\Delta = \tfrac{1}{2} S_{111} u_1 \prod_{s=2}^{N} S_{ss}$$

which shows that the path is stable on one side of the critical point ($u_1 = 0$) and unstable on the other side.

Alternatively, by substituting for u_1 and u_s in the determinant (1.1.63) one gets

$$\Delta = -S'_{11} \lambda \prod_{s=2}^{N} S_{ss} \tag{1.1.65}$$

which when compared with (1.1.64) shows clearly the phenomenon of exchange of stability discussed earlier and illustrated in Figure 1.7.

1.1.7 Stability distribution on the equilibrium paths

For symmetric points of bifurcation, the determinant (1.1.63) takes the form

$$\Delta = \left[S_{11s} u_s + \frac{1}{2}\left(S_{1111} - 2 \sum_{s=2}^{N} \frac{(S_{s11})^2}{S_{ss}} \right) u_1^2 + S'_{11} \lambda \right] \times \prod_{s=2}^{N} S_{ss} + (\cdots) + \cdots$$

Evaluation at an arbitrary state on the symmetric post-buckling path yields to a first approximation

$$\Delta = \frac{1}{3}\left[S_{1111} - 3 \sum_{s=2}^{N} \frac{(S_{s11})^2}{S_{ss}} \right] u_1^2 \prod_{s=2}^{N} S_{ss} \tag{1.1.66}$$

which indicates clearly that the stability of states on the post-buckling path is not dependent on the coordinate u_1. In fact, the sign of the expression in the brackets determines the stability of the initial post-buckling path as a whole so that *if this expression is positive (negative) the path is totally stable (unstable)*. It must also be noted that, when the loss of stability occurs while Λ is increased in the positive direction, normally, $S'_{11} < 0$ and the curvature of the post-buckling path (1.1.59) is, then, positive (negative) when the path is stable (unstable).

Finally, the stability of the critical point itself can be examined by following the procedure given in Section 1.1.4, and observing that the derivatives of H become equivalent to the derivatives of S. It follows that the asymmetric point of bifurcation characterized by $S_{111} \neq 0$ is unstable on the basis of arguments given in Section 1.1.4. In fact, the energy surface has a point of inflexion at C in the u_1 direction due to $S_{111} \neq 0$, ruling out the possibility of a minimum. Similarly, for stable points of bifurcation, one observes that the derivative (1.1.38) deciding on the stability of the critical point is exactly the same as the expression in brackets in (1.1.66) deciding on the stability of the post-buckling path. Hence the following general result can be stated:

The initial post-buckling path is stable or unstable according to whether the critical point itself is stable or unstable, respectively.

This important result was first obtained by Koiter for continuous systems [12].

The following question may, then, be asked: In assessing the post-buckling characteristics of a system, is it always necessary to find the initial post-buckling paths in the load-deflection space or would it be more convenient to examine the stability of the critical point itself? The two problems might be thought to be in a sense identical but there does exist a certain difference in their solution. As illustrated in the preceding sections, the latter approach dispenses with the variable loading parameter in the course of analysis aimed at establishing the convexity of the

1.1 Simple (discrete) critical points

energy surface at the critical point, and may be simpler. It will be seen that the behaviour of the imperfect system also depends on the stability of the critical point itself.

Problem. Examine the stability distribution in the vicinity of a limit point in a similar way. Use the function H and the corresponding determinant.

1.2

Imperfect systems: simple critical points

1.2.1 General remarks

Practical structural systems are likely to have small built-in imperfections in various forms such as unavoidable eccentricities in the loading, geometrical irregularities and material defects. It is natural to suppose that the equilibrium paths of a slightly imperfect version of the same structural system will not lie far away from those of the idealized (perfect) system in the load-deflection space, and attention focuses on the paths in the vicinity of the critical point belonging to the idealized system. Koiter explored the equilibrium paths of the imperfect system in the vicinity of a bifurcation point by introducing the imperfection parameter ϵ into the potential energy functional. He showed that the behaviour of the imperfect system is largely determined by the post-buckling behaviour of the idealized system which is linked to the stability of the bifurcation point itself. Koiter's theory explains when and how the bifurcation load of the idealized system can be reduced drastically even by small imperfections. Koiter's theory concerning imperfections has been verified experimentally in an interesting series of tests by Roorda on plane triangulated frames [36], who has also examined the effect of another type of imperfection which he calls 'minor' [59]. Thompson studied [41] the effect of imperfections through the perturbation technique.

1.2.2 Equilibrium analysis

In this section, the analysis will be concerned with systems which, in their idealized form, lose stability at a bifurcation point. One might enquire about the effect of imperfections on systems losing stability at a limit point. The stability analysis of multiple-parameter systems [43, 45], has shown that such an effect is insignificant. Thus, interpreting one of the

1.2 Imperfect systems: simple critical points

parameters Λ^i as an imperfection parameter, one can immediately observe that the critical relationship has a finite slope and the critical load, therefore, is not sensitive to imperfections (see Section 2.2.3 and the problem at the end of this section).

The formulation in the preceding sections may serve as a basis for the analysis of the imperfect systems. Thus, adding the imperfection parameter ϵ to the variables in the energy function $V(Q_i, \Lambda)$, one obtains the single-valued function

$$V = V(Q_i, \Lambda, \epsilon)$$

which describes a family of imperfect systems, $\epsilon = 0$ giving the idealized system. Following the arguments in Section 1.1.6, the potential energy of the imperfect system can be referred to the fundamental path of the idealized system by setting

$$S(u_i, \Lambda, \epsilon) \equiv V[Q_i(\Lambda) + \alpha_{ij}u_j, \Lambda, \epsilon] \qquad (1.2.1)$$

where the function $S(u_i, \Lambda, 0)$ is related to the perfect system with its only necessary properties (1.1.52) and (1.1.53). This formulation indicates that as ϵ tends to zero, the behaviour of the system described by (1.2.1) approaches asymptotically that of idealized system.

The system now has $N+2$ variables as opposed to N equilibrium equations, and the solutions can be expressed in terms of two independent parameters. Thus, choosing u_1 and Λ as independent parameters the equilibrium paths of the imperfect system in the vicinity of a bifurcation point C can be assumed to have the form

$$u_s = u_s(u_1, \Lambda), \qquad \epsilon = \epsilon(u_1, \Lambda)$$

Substitution of these solutions back into the equilibrium equations results in the identities

$$S_i[u_s(u_1, \Lambda), \epsilon(u_1, \Lambda), u_1, \Lambda] \equiv 0 \qquad (1.2.2)$$

which can, then, be differentiated with respect to the independent variables, u_1 and Λ as many times as required, each step resulting in a differentiated form equal to zero.

Differentiating (1.2.2) once with respect to u_1 and once with respect to Λ yields

$$S_{is}u_{s,1} + \dot{S}_i\epsilon_1 + S_{i1} = 0$$

and $\qquad (1.2.3)$

$$S_{is}u'_s + \dot{S}_i\epsilon' + S'_i = 0$$

respectively, where a dot denotes differentiation with respect to ϵ.

1.2.2 Equilibrium analysis

Evaluation at the critical point C leads to the derivatives

$$\epsilon_1 = 0 \quad \text{for} \quad i = 1$$
$$u_{s,1} = 0 \quad \text{for} \quad i = s \tag{1.2.4}$$

and

$$\epsilon' = 0 \quad \text{for} \quad i = 1$$
$$u'_s = 0 \quad \text{for} \quad i = s \tag{1.2.5}$$

Here, it is tacitly assumed that $\dot{S}_1 \equiv \dot{S}_1(0, \Lambda^c, 0) \neq 0$, specifying the imperfections as 'major' in Roorda's terminology. To define and analyze the effect of 'minor' imperfections, Roorda sets $\dot{S}_1 \equiv 0$.

Differentiate now (1.2.2) with respect to u_1 twice to obtain

$$(S_{isr}u_{r,1} + \dot{S}_{is}\epsilon_1 + S_{is1})u_{s,1} + S_{is}u_{s,11}$$
$$+ (\dot{S}_{is}u_{s,1} + \ddot{S}_i\epsilon_1 + \dot{S}_{i1})\epsilon_1 + \dot{S}_i\epsilon_{11}$$
$$+ S_{i1s}u_{s,1} + \dot{S}_{i1}\epsilon_1 + S_{i11} = 0$$

which, on evaluation at C, yields

$$\epsilon_{11} = -\frac{S_{111}}{\dot{S}_1} \quad \text{for} \quad i = 1$$

and (1.2.6)

$$u_{s,11} = \frac{1}{S_{ss}}\left(\frac{\dot{S}_s}{\dot{S}_1}S_{111} - S_{s11}\right) \quad \text{for} \quad i = s,$$

assuming that $S_{111} \neq 0$ for now.

Similarly, differentiating (1.2.2) with respect to Λ twice yields

$$(\cdots)u'_s + S_{is}u''_s + (\cdots)\epsilon' + \dot{S}_i\epsilon''$$
$$+ S'_{is}u'_s + \dot{S}'_i\epsilon' + S''_i = 0$$

which upon evaluation at C, and of course using (1.2.4) and (1.2.5), results in

$$\epsilon'' = 0 \quad \text{for} \quad i = 1$$

and (1.2.7)

$$u''_s = 0 \quad \text{for} \quad i = s$$

It is understood that the exact expressions in brackets $(\cdots)$ are not needed in view of the fact that u'_s, ϵ', etc. vanish.

Differentiate now (1.2.2) once with respect to u_1 and for a second time with respect to Λ to get

$$(\cdots)u_{s,1} + S_{is}u'_{s,1} + (\cdots)\epsilon_1 + \dot{S}_i\epsilon'_1$$
$$+ S_{i1s}u'_s + \dot{S}_{i1}\epsilon' + S'_{i1} = 0$$

1.2 Imperfect systems: simple critical points

which yields

$$\epsilon'_1 = -\frac{S'_{11}}{\dot{S}_1} \quad \text{and} \quad u'_{s,1} = \frac{S'_{11}}{S_{ss}} \frac{\dot{S}_s}{\dot{S}_1} \tag{1.2.8}$$

The relations $u_s = u_s(u_1, \Lambda)$ and $\epsilon = \epsilon(u_1, \Lambda)$ can now be expressed asymptotically in the vicinity of the asymmetric point of bifurcation. Thus, using the derivatives (1.2.4 to 1.2.8) and introducing $\Lambda = \Lambda^c + \lambda$, Taylor's expansion yields

$$\dot{S}_1 \epsilon + \tfrac{1}{2} S_{111} u_1^2 + S'_{11} u_1 \lambda = 0$$

and (1.2.9)

$$S_{ss} u_s - \frac{1}{2} \left(S_{111} \frac{\dot{S}_s}{\dot{S}_1} - S_{s11} \right) u_1^2 - \frac{\dot{S}_s}{\dot{S}_1} S'_{11} u_1 \lambda = 0,$$

defining a surface in the vicinity of C, which can be considered as a special form of the more general 'equilibrium surface' defined in Ref. [43]. The latter concept will be introduced and treated fully in Part 2.

Attention here is naturally focussed on the projection of this two-dimensional surface into the $u_1 - \lambda - \epsilon$ subspace, defined by the first of the equations (1.2.9). This surface is *anticlastic* and its intersections with the planes $\epsilon = \text{const}$ yield a family of hyperbolae sharing the solution of the idealized system as their asymptotes. In fact, if one sets $\epsilon = 0$ in the first of equations (1.2.9), the fundamental path $u_1 = 0$ and post-buckling path (1.1.58), intersecting at C, are obtained. Figure 1.9 illustrates the paths of

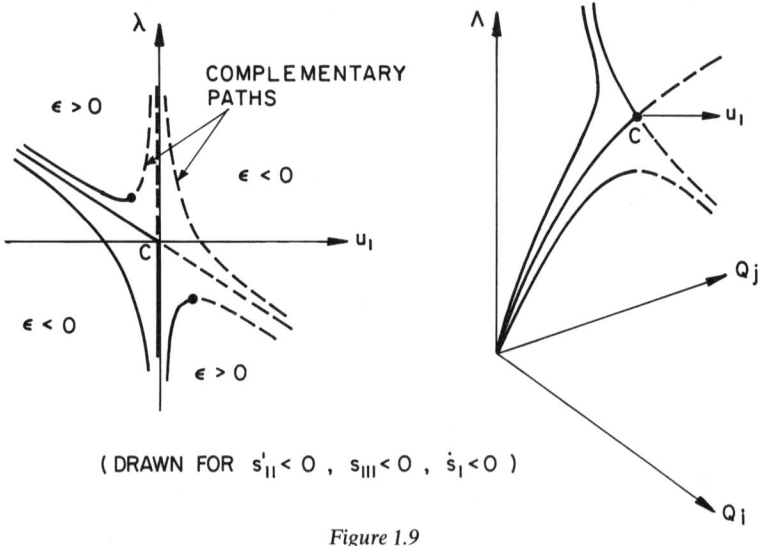

(DRAWN FOR $s'_{11} < 0$, $s_{111} < 0$, $\dot{s}_1 < 0$)

Figure 1.9

1.2.2 Equilibrium analysis

the system in the transformed and original coordinates, only *natural* paths being shown in the latter. It is seen that, for a certain value of the imperfection parameter ϵ, the equilibrium paths of the imperfect system may take two possible forms when the loading parameter is increased from zero (natural paths): either a limit point is reached at a load below the bifurcation point of the idealized system or a stable path rises continuously, instability not being exhibited in the vicinity of C. An interesting feature is the existence of *complementary* paths which cannot occur in a natural process of loading, but nevertheless have been shown to exist also experimentally by Roorda.

The most important aspect of the phenomena described above is, no doubt, the occurrence of a limit point instead of bifurcation at a considerably reduced value of the loading parameter when an appropriate small imperfection is present. The relationship between the critical peak loads and ϵ, therefore, gains significance and will be discussed later on. For a complete picture of paths, however, the behaviour of the imperfect system in the vicinity of a symmetric point of bifurcation must first be analyzed.

Setting $S_{111} = 0$ in (1.2.6) results in a vanishing derivative ϵ_{11}, and one proceeds to determine the next derivative for a first order asymptotic path expression. Thus, differentiating (1.2.2) with respect to u_1 for a third time and evaluating at C for $S_{111} = 0$ yields

$$\epsilon_{111} = -\frac{1}{S_1'}\left(S_{1111} - 3\sum_{s=2}^{N}\frac{(S_{s11})^2}{S_{ss}}\right) \tag{1.2.10}$$

which is used to obtain the first order equation

$$\dot{S}_1 \epsilon + S_{11}' u_1 \lambda + \frac{1}{3!}\left(S_{1111} - 3\sum_{s=2}^{N}\frac{(S_{s11})^2}{S_{ss}}\right)u_1^3 = 0 \tag{1.2.11}$$

in the $u_1 - \lambda - \epsilon$ subspace. Obviously the order of terms is not affected in $u_s = u_s(u_1, \lambda)$ expression which can readily be obtained by setting $S_{111} = 0$ in (1.2.9) as

$$S_{ss}u_s + \tfrac{1}{2}S_{s11}u_1^2 - S_{11}'\frac{\dot{S}_s}{\dot{S}_1}u_1\lambda = 0 \tag{1.2.12}$$

Setting $\epsilon = 0$ in (1.2.11), one obtains the fundamental path $u_1 = 0$ and post-buckling path (1.1.60) of the idealized system. Figures 1.10 and 1.11 illustrate the equilibrium paths of imperfect system in the vicinity of a stable and unstable bifurcation point respectively; these paths are obtained from (1.2.11) by setting $\epsilon = \text{const}$. It is observed that the sign of the imperfection parameter here does not change the nature of the response, and the natural paths of imperfect systems rise continuously and remain

1.2 Imperfect systems: simple critical points

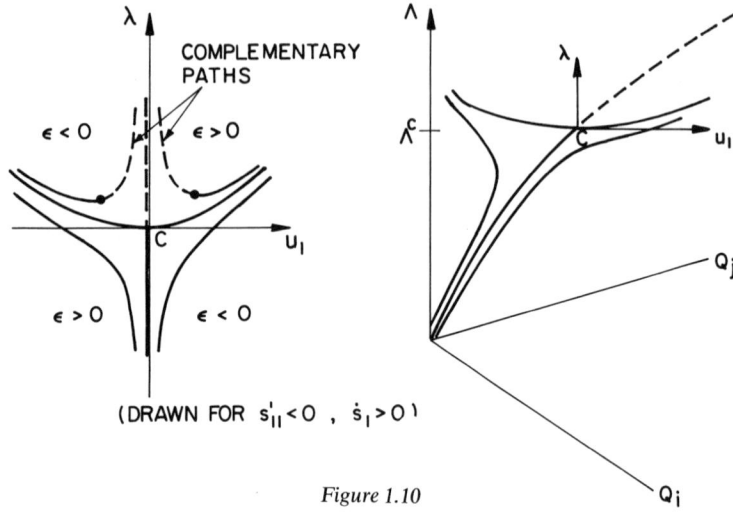

Figure 1.10

stable in the vicinity of a stable-symmetric point of bifurcation, loss of stability being exhibited on the complementary paths which are of academic interest only. In the case of unstable-symmetric points of bifurcation, however, imperfections play a far more significant role, always resulting in a loss of stability possibly at highly reduced values of Λ on reaching a limit point.

Thus, combining the behaviour of the imperfect systems in the vicinity of three bifurcation points, the following result may be stated:

If the bifurcation point itself is unstable the critical loads of the associated impefect systems can be reduced drastically.

Such systems are said to be 'imperfection sensitive'.

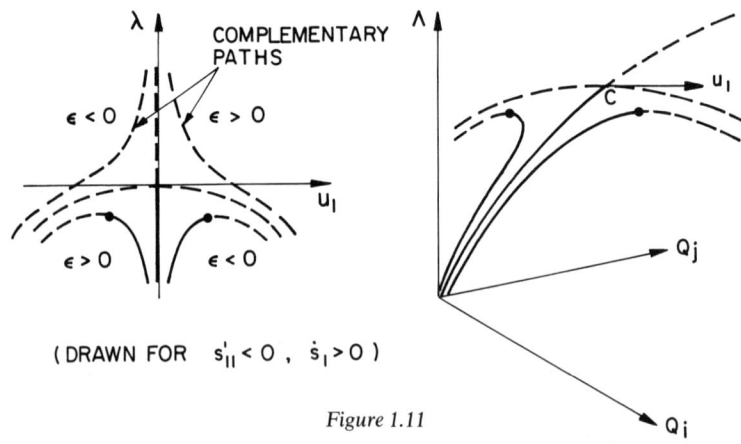

Figure 1.11

1.2.3 Stability analysis: critical load-imperfection relationship

The variation of the critical load with the imperfection parameter ϵ can be examined in two ways, and both will now be demonstrated.

First, it is noted that for all critical points lying on the surface (1.2.9), or (1.2.11) and (1.2.12) in the symmetric case, the stability determinant

$$\Delta(u_k, \Lambda, \epsilon) \equiv \det |S_{ij}(u_k, \Lambda, \epsilon)| \tag{1.2.13}$$

vanishes, and the first order expansion of this determinant can be solved simultaneously with (1.2.9), or (1.2.11) and (1.2.12) to obtain the *critical line* on those surfaces respectively. For this purpose, in addition to the derivatives of the determinant $\Delta(u_k, \Lambda, 0)$ obtained in Section 1.1.7, the derivative $\dot{\Delta} \equiv \dfrac{\partial \Lambda}{\partial \epsilon}\bigg|_c$ is required and can readily be determined as

$$\dot{\Delta} = \dot{S}_{11} \prod_{s=2}^{N} S_{ss} \tag{1.2.14}$$

which is, then, used together with (1.1.62a) to construct the first order determinantal equation

$$S_{11i} u_i + S'_{11} \lambda + \dot{S}_{11} \epsilon = 0 \tag{1.2.15}$$

for asymmetric points of bifurcation.

Solving (1.2.15) and (1.2.9) simultaneously, the first order critical load-imperfection relationships

$$\dot{S}_1 \epsilon - \frac{1}{2} \frac{(S'_{11})^2}{S_{111}} \overset{*}{\lambda}{}^2 = 0 \tag{1.2.16}$$

and

$$S_{111} \overset{*}{u}_1 + S'_{11} \overset{*}{\lambda} = 0 \tag{1.2.17}$$

are obtained. Here a star is used to denote the critical variables. Equations (1.2.16) and (1.2.17) define a *critical zone* on the equilibrium surface (see Section 2.2.3). For symmetric points of bifurcation, first order expansion of the determinant (1.2.13) takes the form

$$S_{11s} u_s + \frac{1}{2}\left(S_{1111} - 2\sum_{s=2}^{N} \frac{(S_{s11})^2}{S_{ss}}\right) u_1^2 + \dot{S}_{11} \epsilon + S'_{11} \lambda = 0 \tag{1.2.18}$$

which, upon solving simultaneously with (1.2.11) and (1.2.12), yields the asymptotic relations

$$S'_{11} \overset{*}{\lambda} + \tfrac{1}{2} \bar{S}_{1111} \overset{*}{u}_1^2 = 0 \tag{1.2.19}$$

1.2 Imperfect systems: simple critical points

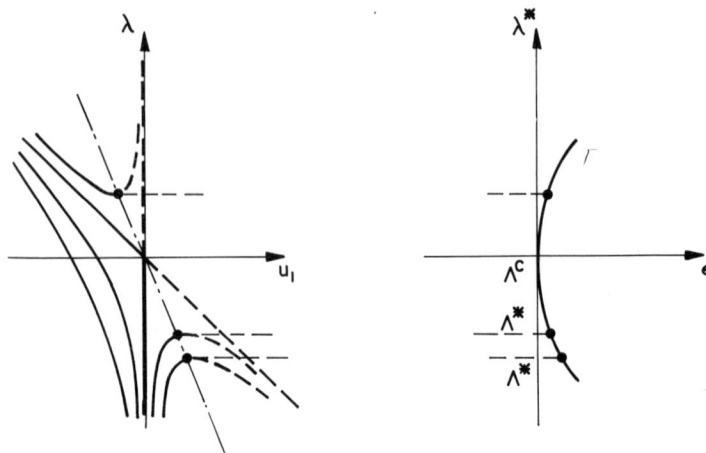

Figure 1.12

and

$$S'_{11}\overset{*}{\lambda} + \tfrac{1}{2}\bar{S}^{\frac{1}{3}}_{1111}(3\dot{S}_1\epsilon)^{\frac{2}{3}} = 0 \tag{1.2.20}$$

where

$$\bar{S}_{1111} = S_{1111} - 3\sum_{s=2}^{N} \frac{(S_{s11})^2}{S_{ss}}$$

Naturally the critical load-imperfection relationship is more important than the other critical relations derived for asymmetric and symmetric points of bifurcation. For the former points, the relation (1.2.16) describes a parabola and is shown in Figure 1.12. It is seen that, due to the infinite slope at Λ^c, even a small imperfection may reduce the critical load of the system considerably; the structure is, then, said to be imperfection-sensitive.

In the vicinity of a symmetric point of bifurcation, the corresponding relation, (1.2.20), is in the form of a cusp illustrated in Figure 1.13. In a natural mode of loading from zero, only unstable-symmetric points exhibit a loss of stability, and as Figure 1.13 shows there is again an extreme sensitivity to initial imperfections. Furthermore, as opposed to asymmetric case, the system now exhibits sensitivity to both positive and negative imperfections.

For completeness, the critical $\overset{*}{\lambda} - \epsilon$ relation in the vicinity of a stable-symmetric point of bifurcation is also shown in Figure 1.14; it is clear, however, that the cusp shown cannot arise in a natural mode of loading from zero and all critical points lie on complementary paths.

The critical relations (1.2.16) and (1.2.20) can also be obtained by

1.2.3 Stability analysis: critical load-imperfection relationship

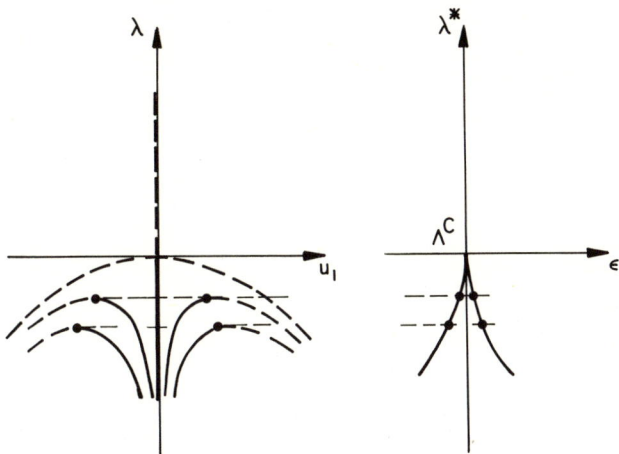

Figure 1.13

applying a more direct perturbation technique, and this second intrinsic approach will now be described.

Critical equilibrium states satisfy N equilibrium equations $S_i = 0$ and $\Delta = 0$; it is, therefore, clear that solutions can be expressed in terms of one parameter as opposed to two-parameter solutions of the preceding section describing the equilibrium surface.

Choosing the functions in the form

$$\overset{*}{u}_s = \overset{*}{u}_s(\overset{*}{u}_1), \qquad \overset{*}{\Lambda} = \overset{*}{\Lambda}(\overset{*}{u}_1), \qquad \epsilon = \epsilon(\overset{*}{u}_1)$$

and substituting back into $S_i = 0$ and $\Delta = 0$, one obtains the conditions of critical equilibrium in the form of identities

$$S_i[\overset{*}{u}_s(\overset{*}{u}_1), \overset{*}{\Lambda}(\overset{*}{u}_1), \epsilon(\overset{*}{u}_1), \overset{*}{u}_1] \equiv 0 \tag{1.2.21}$$

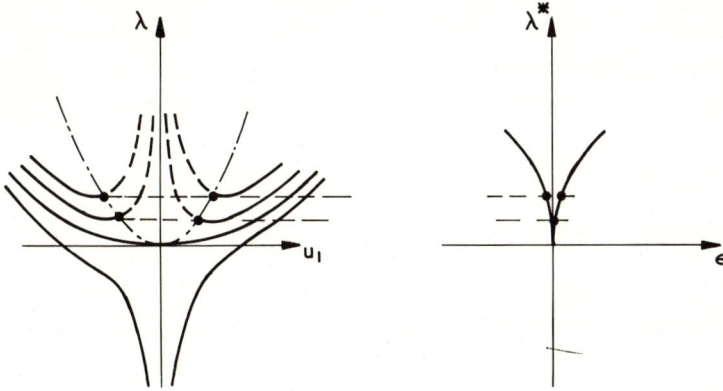

Figure 1.14

1.2 Imperfect systems: simple critical points

and

$$\Delta[\overset{*}{u}_s(\overset{*}{u}_1), \overset{*}{\Lambda}(\overset{*}{u}_1), \epsilon(\overset{*}{u}_1), \overset{*}{u}_1] \equiv 0 \tag{1.2.22}$$

Differentiating these functions with respect to $\overset{*}{u}_1$ results in

$$S_{is}\overset{*}{u}_{s,1} + S'_i\overset{*}{\Lambda}_1 + \dot{S}_i\epsilon_1 + S_{i1} = 0$$

and

$$\Delta_s\overset{*}{u}_{s,1} + \Delta'\overset{*}{\Lambda}_1 + \dot{\Delta}\epsilon_1 + \Delta_1 = 0$$

which upon evaluation at the critical point, given by $\overset{*}{u}_i = 0$, $\overset{*}{\Lambda} = \Lambda^c$ and $\epsilon = 0$, yield the derivatives

$$\epsilon_1 = 0, \quad \overset{*}{u}_{s,1} = 0, \quad \overset{*}{\Lambda}_1 = -\frac{S_{111}}{S'_{11}} \tag{1.2.23}$$

It is understood that the derivatives (1.1.62a) are used to obtain (1.2.23).

A second differentiation of the identities yields

$$(\cdots)\overset{*}{u}_{s,1} + S_{is}\overset{*}{u}_{s,11} + (S'_{is}\overset{*}{u}_{s,1} + S''_i\overset{*}{\Lambda}_1 + \dot{S}'_i\epsilon_1 + S'_{i1})\overset{*}{\Lambda}_1$$
$$+ S'_i\overset{*}{\Lambda}_{11} + (\cdots)\epsilon_1 + \dot{S}_i\epsilon_{11} + S_{i1s}\overset{*}{u}_{s,1} + S'_{i1}\overset{*}{\Lambda}_1$$
$$+ S_{i1}\epsilon_1 + S_{i11} = 0$$

and

$$(\cdots)\overset{*}{u}_{s,1} + \Delta_s\overset{*}{u}_{s,11} + (\Delta'_s\overset{*}{u}_{s,1} + \Delta''\overset{*}{\Lambda}_1 + \dot{\Delta}'\epsilon_1 + \Delta'_1)\overset{*}{\Lambda}_1$$
$$+ \Delta'\overset{*}{\Lambda}_{11} + (\cdots)\epsilon_1 + \dot{\Delta}\epsilon_{11} + \Delta_{1s}\overset{*}{u}_{s,1} + \Delta'_1\overset{*}{\Lambda}_1$$
$$+ \dot{\Delta}_1\epsilon_1 + \Delta_{11} = 0$$

which upon evaluation at the critical point results in the derivatives

$$\epsilon_{11} = \frac{S_{111}}{\dot{S}_1}, \quad \overset{*}{u}_{s,11} = -\frac{1}{S_{ss}}\left(\frac{\dot{S}_s}{\dot{S}_1}S_{111} + S_{s11}\right) \tag{1.2.24}$$

and, for symmetric points of bifurcation where S_{111} and consequently $\overset{*}{\Lambda}_1$ vanish,

$$\overset{*}{\Lambda}_{11} = -\frac{1}{S'_{11}}\left(S_{1111} - 3\sum_{s=2}^{N}\frac{(S_{s11})^2}{S_{ss}}\right) \tag{1.2.25}$$

It is further noted that at a symmetric point of bifurcation $\epsilon_{11} = 0$, and one proceeds to evaluate the higher order derivatives as may be required. Differentiating the identities with respect to $\overset{*}{u}_1$ for a third time and evaluating at C results in

$$\epsilon_{111} = \frac{2}{\dot{S}}\left(S_{1111} - 3\sum_{s=2}^{N}\frac{(S_{s11})^2}{S_{ss}}\right) \tag{1.2.26}$$

1.2.3 Stability analysis: critical load-imperfection relationship

From (1.2.23 to 1.2.26), the critical relations obtained before can now be reconstructed. Thus, using Taylor's expansion, the relations (1.2.16) and (1.2.17) for asymmetric points, and the relations (1.2.19) and (1.2.20) for symmetric points are recovered.

It may be useful to have the critical load-imperfection relationships in terms of $\overset{*}{\Lambda} - \epsilon$ rather than $\overset{*}{\lambda} - \epsilon$. This can readily be achieved by substituting $\overset{*}{\Lambda} = \Lambda^c + \overset{*}{\lambda}$ into these equations, resulting in

$$\overset{*}{\Lambda} = \Lambda^c \pm \frac{(2\dot{S}_1 S_{111} \epsilon)^{\frac{1}{2}}}{S'_{11}} \qquad (1.2.27)$$

for asymmetric bifurcation points, and

$$\overset{*}{\Lambda} = \Lambda^c - \tfrac{1}{2}\bar{S}_{1111}^{\frac{1}{3}} \frac{(3\dot{S}_1 \epsilon)^{\frac{2}{3}}}{S'_{11}} \qquad (1.2.28)$$

for symmetric points of bifurcation.

One advantage of the latter procedure is its directness, reducing substitutions. Furthermore, each step of ordered perturbations produces a set of derivatives which might reveal certain important features of the behaviour, without necessitating higher order perturbations. For example, comparing (1.1.57) and (1.2.23) without performing the rest of the analysis, one observes that

$$\overset{*}{\Lambda}_1 = 2\Lambda$$

for asymmetric points, and similarly from (1.1.59) and (1.2.25)

$$\overset{*}{\Lambda}_{11} = 3\Lambda_{11}$$

for symmetric points, which indicate the general orientation of the *critical line* with respect to post-buckling paths. The dash-dotted lines in Figures 1.12, 1.13 and 1.14 show the critical lines.

Problem. Introducing the imperfection parameter ϵ into the energy function H, perform an analysis similar to the preceding one to obtain the critical load-imperfection relationship in the vicinity of a limit point. Show that $\partial \overset{*}{\lambda}/\partial \epsilon$ is, in general, finite and the system is not, therefore, imperfection-sensitive.

1.3

Symmetric systems: simple critical points

1.3.1 General remarks

Many structural systems exhibit certain symmetry properties in the instability behaviour, and such a phenomenon has already been discussed in connection with the vanishing cubic coefficient S_{111}. It must immediately be remarked here that the physical symmetry of a system does not necessarily yield a symmetric point of bifurcation. In fact, a symmetric structure which is loaded symmetrically can very well exhibit an asymmetric point of bifurcation, and experiments performed on a symmetrically loaded bridge truss [59] supply quantitative evidence of this type of behaviour. The explanation for this apparently curious phenomenon must be sought in the composition of the total potential energy function. In the problem of the bridge truss, for example, the energy function is not symmetric with respect to buckling modes, and the cubic energy coefficient does not vanish. In other words, positive and negative buckling modes of a member in the truss do not possess the property of mirror symmetry and the associated energies are not the same.

On the other hand, a significant class of structural systems exhibits symmetry in the buckling behaviour as well as in the composition of the total potential energy function. In developing a general theory for such systems a *mathematical model* can conveniently be produced by simply introducing appropriate symmetry conditions into the potential energy function. For discrete systems, this is achieved by separating the generalized coordinates into two distinct sets and assuming that the potential energy function is symmetric in one of the sets [60, 61, 43, 48, 62]. Such a system will then exhibit symmetry in the buckling behaviour. In this regard, Masur [63] recently discussed completely symmetric structures in the continuum context.

1.3 Symmetric systems: simple critical points

1.3.2 Potential energy and symmetry

Consider the structural system characterized by the total potential energy function

$$V = V(Q_i, z_\alpha, \Lambda), \quad \begin{aligned} i &= 1, 2, \ldots, N; \\ \alpha &= 1, 2, \ldots, K \end{aligned} \quad (1.3.1)$$

which is assumed to be single-valued and continuously differentiable at least in the region of interest. Generalized coordinates now consist of two distinct sets represented by Q_i and z_α. It is assumed that these two sets of coordinates describe two different types of deformations, and the function V is symmetric in the z_α such that

$$V(Q_i, z_\alpha, \Lambda) = V(Q_i, -z_\alpha, \Lambda) \quad (1.3.2)$$

where the z_α *change sign as a set*. Consequently, in the expansion of the potential energy function about the origin, terms such as

$$z_\alpha, z_\alpha z_\beta z_\gamma, \ldots, Q_i z_\alpha, \ldots, \Lambda z_\alpha \cdots$$

cannot appear, and it can readily be verified that this property holds for expansion about any point on the fundamental path.

Various buckling problems of plates form typical examples of this class of systems, Q_i and z_α representing *in-plane* and *out-of-plane* (buckling) deformations respectively. Symmetrically loaded shallow arches and shallow spherical shells exhibit similar symmetry properties such that the set Q_i describes symmetric deformations while the set z_α represents the asymmetric (buckling) deformations.

It was noted in Section 1.1.2 that the total potential energy function of most structural systems is linear in the loading parameter Λ as expressed in the form (1.1.2), and the analysis here will be based on this form. Thus, corresponding to (1.1.2) one has

$$V = U(Q_i, z_\alpha) - \Lambda E(Q_i, z_\alpha) \quad (1.3.3)$$

Expanding the strain energy function $U(Q_i, z_\alpha)$ and the generalized deflection function $E(Q_i, z_\alpha)$ into Taylor's series about the origin (zero-state), remembering that this point is one of equilibrium, and using the symmetry properties results in

$$\begin{aligned} V = &\frac{1}{2!}(U_{ij}Q_iQ_j + U_{\alpha\beta}z_\alpha z_\beta) \\ &+ \frac{1}{3!}(U_{ijk}Q_iQ_jQ_k + 3U_{i\alpha\beta}Q_i z_\alpha z_\beta) \\ &+ \frac{1}{4!}(U_{ijkl}Q_iQ_jQ_kQ_l + 6U_{ij\alpha\beta}Q_iQ_j z_\alpha z_\beta + U_{\alpha\beta\gamma\delta}z_\alpha z_\beta z_\gamma z_\delta) \\ &+ \frac{1}{5!}(\cdots) + \cdots - \Lambda E_i Q_i \end{aligned} \quad (1.3.4)$$

1.3.2 Potential energy and symmetry

Here, the notation is self-evident and the summation convention as before is employed. In view of the fact that deflections are described by two distinct sets of generalized coordinates, the function E is assumed to be linear in the generalized coordinates (as it will be in specific problems treated appropriately), the z_i thus not being involved due to symmetry properties.

A comparison of the strain energy expression (1.3.4) with the large deflection energy functional of plates given by Timoshenko [64] reveals that the terms involved in both expressions are similar providing the third and fourth order terms in the Q_i and the fourth mixed term in (1.3.4) are omitted. It is, then, observed that the displacements u and v (in-plane) in Timoshenko's expression correspond to the Q_i while w (out-of-plane) is represented by the z_α. In this reduced special form, (1.3.4) can, therefore, be considered as the large-deflection energy expression of a discretized plate.

$N + K$ equilibrium equations can now be expressed in the form of two distinct sets, $V_i = 0$ $(i = 1, 2, \ldots, N)$ and $V_\alpha = 0$ $(\alpha = 1, 2, \ldots, K)$ which are given as

$$\frac{\partial V}{\partial Q_i} = U_{ij}Q_j + \frac{1}{2!} U_{ijk}Q_jQ_k + \cdots$$

$$+ \frac{1}{2!} U_{i\alpha\beta}z_\alpha z_\beta + \cdots - \Lambda E_i = 0 \tag{1.3.5}$$

$$\frac{\partial V}{\partial z_\alpha} = U_{\alpha\beta}z_\beta + U_{i\alpha\beta}Q_i z_\beta + \cdots = 0 \tag{1.3.6}$$

One obvious solution of these equations is easily obtained as

$$z_\alpha = 0$$
$$U_{ij}Q_j + \tfrac{1}{2} U_{ijk}Q_jQ_k + \cdots - \Lambda E_i = 0 \tag{1.3.7}$$

which defines a fundamental equilibrium path in the load-deflection space, in fact, in the $Q_i - \Lambda$ sub-space. Evidently, the Q_i and z_α represent two different types of deflections, the latter being *superimposed* on the former (Figure 1.15). In other words, systems under consideration start deflecting symmetrically as Λ is increased gradually from zero, and a generally nonlinear fundamental equilibrium path emerges from the origin and develops entirely in the $Q_i - \Lambda$ subspace until a critical bifurcation point is reached (if it exists) at which asymmetric buckling deflections (the z_α) come into the picture causing buckling. Depending on the physical properties of a given system, however, the fundamental path may lose stability on reaching an extremum, the latter deflections not developing prior to such a limit point.

1.3 Symmetric systems: simple critical points

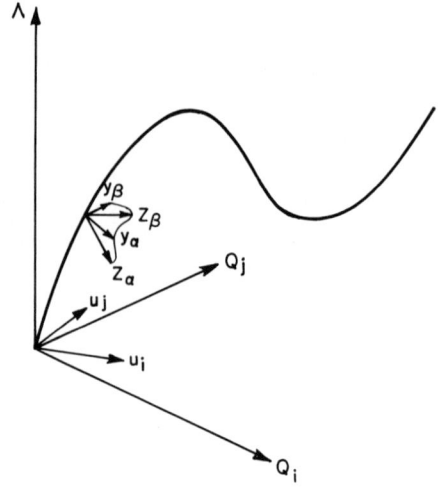

Figure 1.15

In the next two sections, ordered estimates of the fundamental path, bifurcation and limit loads will be obtained through perturbations, analysis following a similar pattern to that of Section 1.1.5; here, however, results will take an explicit form.

The following analyses will be facilitated by introducing the orthogonal transformations

$$Q_i = a_{ij} u_j, \qquad a_{ij} a_{jk} = \delta_{ik}$$

and (1.3.8)

$$z_\alpha = b_{\alpha\beta} y_\beta, \qquad b_{\alpha\beta} b_{\beta\gamma} = \delta_{\alpha\gamma}$$

to diagonalize the quadratic forms of the energy function in the Q_i and z_α respectively. The transformed potential energy function

$$W(u_i, y_\alpha, \Lambda) \equiv U(a_{ij} u_j, b_{\alpha\beta} y_\beta) - \Lambda E(a_{ij} u_j, b_{\alpha\beta} y_\beta) \qquad (1.3.9)$$

now has the following properties:

$$W_{ij} = 0 \text{ for } i \neq j \quad \text{and} \quad W_{\alpha\beta} = 0 \text{ for } \alpha \neq \beta$$

due to (1.3.8), (1.3.10)

$$W_\alpha = W_{\alpha\beta\gamma} = \cdots W_{i\alpha} = \cdots = W'_\alpha = \cdots = 0$$

due to symmetry conditions which are evidently carried over to the function W, and

$$W'_{ij} = W'_{ijk} = \cdots W''_i = \cdots W'''_{ij} \cdots = 0 \qquad (1.3.11)$$

1.3.3 Equilibrium analysis

by virtue of the formulation (1.3.9) and (1.3.4). It is further noted that any combined differentiation with respect to Λ and z_α leads to zero.

1.3.3 Equilibrium analysis

Suppose the fundamental path (1.3.7) is described in the parametric form

$$u_i = u_i(\eta)$$
$$\Lambda = \Lambda(\eta) \tag{1.3.12}$$

Both u_1 and Λ will be given the role of η in the following perturbation procedure aimed at expressing the fundamental path in terms of one independent parameter. First, set $\eta = u_1$ in (1.3.12) and substitute back into the equilibrium equations $W_i = 0$ to obtain

$$W_i[u_j(u_1), \Lambda(u_1)] \equiv 0 \tag{1.3.13}$$

where $u_1(u_1)$ is obviously u_1 and it is not separated from the u_s for the sake of tidiness, now that the reader is familiar with the procedure. Since $z_\alpha = 0$ along the fundamental path, the analysis is confined to $u_i - \Lambda$ subspace as (1.3.13) indicates.

First and second perturbations of (1.3.13) yield

$$W_{ij}u_{j,1} + W'_i\Lambda_1 = 0$$

and

$$(W_{ijk}u_{k,1} + W'_{ij}\Lambda_1)u_{j,1} + W_{ij}u_{j,11}$$
$$+ (W'_{ij}u_{j,1} + W''_i\Lambda_1)\Lambda_1 + W'_i\Lambda_{11} = 0$$

respectively, which upon evaluation at the origin and using the essential properties of the function W result in

$$\Lambda_1 = -\frac{W_{11}}{W'_1}, \quad u_{s,1} = -\frac{W'_s}{W_{ss}}\Lambda_1 = \frac{W_{11}}{W_{ss}}\frac{W'_s}{W'_1}$$

and $\tag{1.3.14}$

$$\Lambda_{11} = -\frac{1}{W'_1} W_{1ij}u_{i,1}u_{j,1},$$

$$u_{s,11} = -\frac{1}{W_{ss}}(W'_s\Lambda_{11} + W_{sij}u_{i,1}u_{j,1})$$

It is noted that each derivative contains the derivatives determined in the preceding step, and that $u_{1,1} = 1$, $u_{1,11} = 0$, etc.

1.3 Symmetric systems: simple critical points

A third perturbation of (1.3.13) yields

$$\Lambda_{111} = -\frac{1}{W_1'}[W_{1ijk}u_{i,1}u_{j,1}u_{k,1} + \tfrac{3}{2}W_{1ij}(u_{i,1}u_{j,11} + u_{i,11}u_{j,1})]$$

and (1.3.15)

$$u_{s,111} = -\frac{1}{W_{ss}}[W_s'\Lambda_{111} + W_{sijk}u_{i,1}u_{j,1}u_{k,1}$$

$$+ \tfrac{3}{2}W_{sij}(u_{i,1}u_{j,11} + u_{i,11}u_{j,1})]$$

Continuing perturbations of (1.3.13) yield higher order derivatives sequentially which are, then, used to construct the fundamental path in the series form

$$u_i = u_{i,1}u_1 + \frac{1}{2!}u_{i,11}u_1^2 + \frac{1}{3!}u_{i,111}u_1^3 + \cdots$$

$$\Lambda = \Lambda_1 u_1 + \frac{1}{2!}\Lambda_{11}u_1^2 + \frac{1}{3!}\Lambda_{111}u_1^3 + \cdots$$
(1.3.16)

Truncating the right-hand sides of equations (1.3.16) after the first, second, etc. terms, one obtains the first, second, etc. estimates of the fundamental path.

Before proceeding to stability analysis, consider now the alternative perturbation procedure in which Λ takes on the role of η. The defining identities are, then, in the form

$$W_i[u_j(\Lambda), \Lambda] \equiv 0 \qquad (1.3.17)$$

which can be differentiated successively to yield

$$W_{ij}u_j' + W_i' = 0,$$

$$(W_{ijk}u_k' + W_{ij}')u_j' + W_{ij}u_j'' + W_{ij}'u_j' + W_i'' = 0,$$

etc., which upon evaluation result in

$$u_i' = -\frac{W_i'}{W_{ii}}, \qquad u_i'' = -\frac{W_{ijk}}{W_{ii}}u_j'u_k', \text{ etc.} \qquad (1.3.18)$$

Using these derivatives, the fundamental path is expressed in the form

$$u_i = u_i'\Lambda + \frac{1}{2!}u_i''\Lambda^2 + \frac{1}{3!}u_i'''\Lambda^3 + \cdots \qquad (1.3.19)$$

Evidently, the choice of Λ as the perturbation parameter η yields an easier symmetric treatment, however, the convergence of the series will not be good in many nonlinear problems.

1.3.4 Stability analysis: critical loads

In order to examine the stability of the fundamental path (1.3.12), the variation of the stability determinant along this path is considered. Since the mixed second derivatives of W, like $\partial^2 W/\partial u_i\, \partial y_\alpha$, evaluated on the fundamental path ($y_\alpha = 0$) vanish due to symmetry properties, the stability determinant is the product of two determinants corresponding to the distinct sets of coordinates, and can be expressed as

$$\Delta(u_i) \equiv \Delta^u(u_i) \cdot \Delta^y(u_i) \equiv \left| \frac{\partial^2 w}{\partial u_i\, \partial u_j} \right| \cdot \left| \frac{\partial^2 w}{\partial y_\alpha\, \partial y_\beta} \right|, \qquad (1.3.20)$$

both determinants being a function of u_i only.

This is an important feature of the class of systems under consideration which will contribute significantly toward a comparatively simpler yet explicit stability analysis. The order of the stability equation required for a prescribed estimate of the critical load, for example, is reduced by one half as compared to the order of the equation which has to be maintained for the same estimate if the analysis does not make use of this feature.

Thus, assuming that a loss of stability will always be associated with a simple (discrete) critical point, one observes that either the determinant Δ^u or Δ^y vanishes when the initially stable fundamental path reaches a critical state as the loading parameter is increased gradually. The determinant

$$\Delta^u(u_k) \equiv |W_{ij}(u_k)| \qquad (1.3.21)$$

vanishes at a limit point on reaching an extremum on the path after symmetric deflections increase continuously with Λ.

Setting $\eta = u_1$ in (1.3.12) and substituting into (1.3.21) one has

$$A^u(u_1) \equiv \Delta^u[u_i(u_1)] \equiv \det |W_{ij}[u_k(u_1)]| \qquad (1.3.22)$$

which immediately yields

$$A_0^u \equiv A^u(0) = \prod_{i=1}^{N} W_{ii} \qquad (1.3.23)$$

Differentiating (1.3.22) with respect to u_1 and evaluating at the origin leads to

$$A_1^u = \Delta_i^u u_{i,1} = \frac{\prod_{k=1}^{N} W_{kk}}{W_{ii}} W_{iij} u_{j,1} \qquad (1.3.24)$$

Here the Δ_i^u are obtained by differentiating (1.3.21) and evaluating at the origin (Problem 1). $u_{j,1}$ are the path derivatives determined in the

1.3 Symmetric systems: simple critical points

equilibrium analysis and are given by (1.3.14). It is noted that summation is implied over i as well as j.

The second differentiation of (1.3.22) yields upon evaluation

$$A_{11}^u = \frac{\prod_{m=1}^{N} W_{mm}}{W_{ii}} \Bigg[(W_{iij} u_{j,11} + W_{iijk} u_{j,1} u_{k,1})$$
$$+ \frac{1}{W_{jj}} (W_{iik} W_{jjl} - W_{ijk} W_{jil}) u_{k,1} u_{l,1} \Bigg] \quad (1.3.25)$$

where in the last parenthesis $i \neq j$.

Again, in (1.3.25), the path derivatives $u_{i,1}$, $u_{i,11}$ are already known from the equilibrium analysis and can directly be substituted from (1.3.14). The verification of (1.3.25) is left as an exercise (Problem 2) for the reader.

Having determined A_0^u, A_1^u, etc., one has the first, second, etc. order stability equations

$$A_0^u + A_1^u u_1 = 0$$
$$A_0^u + A_1^u u_1 + \frac{1}{2!} A_{11}^u u_1^2 = 0 \quad (1.3.26)$$

etc.

which yield an ordered form of estimates for the critical value of u_1. The second equation in (1.3.16) can, then be used to obtain corresponding estimates of the critical value of Λ. Here, a question arises as to the order of approximation to be maintained in this latter equation. It might be thought that for a first order stability estimate, one uses the first order equilibrium equation which yields a straight line. However, it is not physically conceivable to seek limit points on a straight line, and the estimate can, indeed, be improved greatly by using the second order equilibrium equation for a first order stability estimate. This is, of course, due to the fact that only one component of the uncoupled stability determinant is used here.

It is further noted that estimates of limit loads can also be obtained from the equilibrium analysis directly since the limit points represent the extrema on the fundamental path. Such estimates will generally be different from the series of estimates obtained above.

Now suppose physical properties of the system dictate that the determinant

$$\Delta^y(u_i) \equiv \det |W_{\alpha\beta}(u_i)| \quad (1.3.27)$$

vanishes before Δ^u. Then, proceeding along the initially stable fundamental path, a point of bifurcation is reached at which y-type (asymmetric) deflections develop, causing buckling in the asymmetric mode.

1.3.4 Stability analysis: critical loads

Setting again $\eta = u_1$ in (1.3.12) and substituting into (1.3.27), one has

$$A^y(u_1) \equiv \Delta^y[u_i(u_1)] \equiv \det |W_{\alpha\beta}[u_i(u_1)]| \qquad (1.3.28)$$

Proceeding as before one obtains

$$A_0^y = A^y(0) = \prod_{\alpha=1}^{K} W_{\alpha\alpha}$$

$$A_1^y = \frac{\prod_{\beta=1}^{K} W_{\beta\beta}}{W_{\alpha\alpha}} W_{i\alpha\alpha} u_{i,1}, \qquad (1.3.29)$$

$$A_{11}^y = \frac{\prod_{\gamma=1}^{K} W_{\gamma\gamma}}{W_{\alpha\alpha}} \left[(W_{i\alpha\alpha} u_{i,11} + W_{ij\alpha\alpha} u_{i,1} u_{j,1}) \right.$$

$$\left. + \frac{1}{W_{\beta\beta}} (W_{i\alpha\alpha} W_{j\beta\beta} - W_{i\alpha\beta} W_{j\beta\alpha}) u_{i,1} u_{j,1} \right]$$

etc.

where in the last parenthesis $\alpha \neq \beta$.

Ordered stability equations can now be constructed as

$$A_0^y + A_1^y u_1 = 0,$$

$$A_0^y + A_1^y u_1 + \frac{1}{2!} A_{11}^y u_1^2 = 0 \qquad (1.3.30)$$

etc.

which yield ordered estimates of the critical value of u_1. These estimates are, then, used to obtain corresponding estimates of the critical value of Λ from (1.3.16).

Finally, choose Λ as the independent perturbation parameter in the bifurcation case in which the determinant (1.3.27) vanishes. Proceeding as before, the identity

$$A(\Lambda) \equiv \Delta^y[u_i(\Lambda)] \equiv \det |W_{\alpha\beta}[u_i(\Lambda)]| \qquad (1.3.31)$$

is perturbed successively to yield

$$(A^y)' \equiv \left. \frac{dA^y}{d\Lambda} \right|_0 = \frac{\prod_{\gamma=1}^{K} W_{\gamma\gamma}}{W_{\alpha\alpha}} W_{i\alpha\alpha} u_i',$$

$$(A^y)'' = \frac{\prod_{\gamma=1}^{K} W_{\gamma\gamma}}{W_{\alpha\alpha}} \left[(W_{i\alpha\alpha} u_i'' + W_{ij\alpha\alpha} u_i' u_j') \right. \qquad (1.3.32)$$

$$\left. + \frac{1}{W_{\beta\beta}} (W_{i\alpha\alpha} W_{j\beta\beta} - W_{i\alpha\beta} W_{j\beta\alpha}) u_i' u_j' \right]$$

etc.

1.3 Symmetric systems: simple critical points

which are used to construct

$$A_0^y + (A^y)'\Lambda + \frac{1}{2!}(A^y)''\Lambda^2 + \cdots = 0 \tag{1.3.34}$$

Truncations of the left-hand side of this equation after the second, third, etc. terms result in the first, second, etc. estimates of Λ_{cr} directly.

It is seen that the results of this section are in the form of formulae suitable to be used directly in estimating the stability limit of many particular symmetric systems without necessarily repeating the entire analysis, and a specific application will be found in Section 1.4.5 where a shallow circular arch is analyzed.

Problem 1. Choosing u_1 as the perturbation parameter, verify the expressions (1.3.24) and (1.3.25).

Problem 2. Choosing Λ as the perturbation parameter, verify the expressions (1.3.32).

1.3.5 Post-critical behaviour and imperfection sensitivity

The highly nonlinear character of the fundamental path associated with some symmetric structural systems presents certain difficulties in the post-critical analysis. In such cases, the function S which was the basis of the general post-buckling discussion in Section 1.1.6 might prove inappropriate since it involves sliding generalized coordinates along a single-valued fundamental path $Q_i = Q_i^F(\Lambda)$.

In the typical problem of a shallow arch, however, the fundamental path is not single-valued, and in addition, an asymmetric mode of buckling is related to two points of bifurcation on the nonlinear path, resulting in a connected path pattern between the two points.

Although the sliding coordinates may still be used to discuss the behaviour in the vicinity of one of the critical points, an alternative post-buckling analysis, dispensing with such coordinates will now be developed for a specialized system. A discussion concerning the associated imperfection-sensitivities will follow the post-buckling analysis. A more general treatment of post-buckling of symmetric systems is deferred to Section 2.4.4.

Consider a two-degree-of-freedom system described by the total potential energy function

$$W = W(u_1, y_2, \Lambda) \tag{1.3.35}$$

which has the essential properties (1.3.10) and (1.3.11).

Let a prospective post-buckling path passing through the bifurcation

1.3.5 Post-critical behaviour and imperfection sensitivity

point of interest be in the form

$$u_1 = u_1(y_2), \qquad \Lambda = \Lambda(y_2) \tag{1.3.36}$$

such that for $y_2 = 0$, (1.3.36) gives the bifurcation point $u_1 = u_1^B$, $\Lambda = \Lambda^B$ where $W_{22} = 0$.

Substituting (1.3.36) into the equilibrium equations, one has

$$W_1[u_1(y_2), \Lambda(y_2), y_2] \equiv 0 \tag{1.3.37}$$

$$W_2[u_1(y_2), \Lambda(y_2), y_2] \equiv 0 \tag{1.3.38}$$

Differentiating with respect to y_2 yields

$$W_{11} u_{1,2} + W_1' \Lambda_2 + W_{12} = 0 \tag{1.3.39}$$

and

$$W_{21} u_{1,2} + W_2' \Lambda_2 + W_{22} = 0 \tag{1.3.40}$$

Upon evaluation at the bifurcation point (not at the origin), (1.3.39) takes the form

$$W_{11} u_{1,2} + W_1' \Lambda_2 = 0 \tag{1.3.41}$$

while (1.3.40) is identically satisfied.

A second perturbation of (1.3.37) and (1.3.38) yields

$$(W_{111} u_{1,2} + W_{11}' \Lambda_2 + W_{112}) u_{1,2} + W_{11} u_{1,22}$$
$$+ (W_{11}' u_{1,2} + W_1'' \Lambda_2 + W_{12}') \Lambda_2 + W_1' \Lambda_{22}$$
$$+ W_{121} u_{1,2} + W_{12}' \Lambda_2 + W_{122} = 0 \tag{1.3.42}$$

and

$$(W_{211} u_{1,2} + W_{21}' \Lambda_2 + W_{212}) u_{1,2} + W_{21} u_{1,22}$$
$$+ (W_{11}' u_{1,2} + W_2'' \Lambda_2 + W_{22}') \Lambda_2 + W_2' \Lambda_{22}$$
$$+ W_{221} u_{1,2} + W_{22}' \Lambda_2 + W_{222} = 0 \tag{1.3.43}$$

(1.3.43) results in, upon evaluation,

$$u_{1,2} = 0 \tag{1.3.44}$$

Evaluating (1.3.42) and using (1.3.44) one obtains

$$W_{11} u_{1,22} + W_1' \Lambda_{22} + W_{221} = 0 \tag{1.3.45}$$

From (1.3.44) and (1.3.41), it is deduced that

$$\Lambda_2 = 0 \tag{1.3.46}$$

Finally, a third perturbation of (1.3.38) leads to

$$u_{1,22} = -\frac{W_{2222}}{3 W_{221}} \tag{1.3.47}$$

1.3 Symmetric systems: simple critical points

which, in conjunction with (1.3.45), results in

$$\Lambda_{22} = \frac{1}{W'_1}\left(\frac{W_{11}W_{2222}}{3W_{221}} - W_{221}\right) \tag{1.3.48}$$

Using these derivatives, one obtains the asymptotic equations of the initial post-buckling path as

$$u_1 = u_1^B - \frac{1}{6}\frac{W_{2222}}{W_{221}}y_2^2 \tag{1.3.49}$$

$$\Lambda = \Lambda^B + \frac{1}{6W'_1}\left(\frac{W_{11}}{3W_{221}}W_{2222} - W_{221}\right)y_2^2 \tag{1.3.50}$$

The critical values u_1^B and Λ^B are of course available from the stability analysis described in the preceding section.

Imperfect system

The effect of a small imperfection on the behaviour of the system can be studied similarly. Thus, assuming that the *critical line* in the load-deflection space is in the form

$$\overset{*}{u}_1 = \overset{*}{u}_1(y_2), \quad \overset{*}{\Lambda} = \overset{*}{\Lambda}(\overset{*}{y}_2), \quad \epsilon = \epsilon(\overset{*}{y}_2), \tag{1.3.51}$$

and substituting this assumed solution into the conditions of critical equilibrium one has

$$W_1[\overset{*}{u}_1(\overset{*}{y}_2), \overset{*}{\Lambda}(\overset{*}{y}_2), \epsilon(\overset{*}{y}_2), \overset{*}{y}_2] \equiv 0 \tag{1.3.52}$$

$$W_2[\overset{*}{u}_1(\overset{*}{y}_2), \overset{*}{\Lambda}(\overset{*}{y}_2), \epsilon(\overset{*}{y}_2), \overset{*}{y}_2] \equiv 0 \tag{1.3.53}$$

$$\Delta[\overset{*}{u}_1(\overset{*}{y}_2), \epsilon(\overset{*}{y}_2), \overset{*}{y}_2] \equiv 0 \tag{1.3.54}$$

It is interesting to note that, in contrast with the analysis in Section 1.2.3, the stability determinant here is not a function of Λ; this is, of course, due to the essential properties of the function W.

In the following perturbation equations derivatives of Δ will appear, and one proceeds first to evaluate these derivatives. Differentiating and evaluating (at the bifurcation point B) the determinant

$$\Delta(u_1, y_2, \epsilon) = \begin{vmatrix} W_{11}(u_1, y_2, \epsilon) & W_{12}(u_1, y_2, \epsilon) \\ W_{21}(u_1, y_2, \epsilon) & W_{22}(u_1, y_2, \epsilon) \end{vmatrix}$$

one has

$$\Delta_1 = W_{11}W_{221}, \quad \Delta_2 = 0,$$

$$\dot{\Delta} \equiv \frac{\partial \Delta}{\partial \epsilon}\bigg|_B = W_{11}\dot{W}_{22}$$

$$\Delta_{22} = W_{11}W_{2222} - 2(W_{221})^2 \tag{1.3.55}$$

$$\dot{\Delta}_2 = 0$$

1.3.5 Post-critical behaviour and imperfection sensitivity

Turning now to the perturbation analysis, differentiate the identities (1.3.52), (1.3.53) and (1.3.54) to get

$$W_{11}\overset{*}{u}_{1,2} + W_1'\overset{*}{\Lambda}_2 + \dot{W}_1\epsilon_2 + W_{12} = 0$$
$$W_{21}\overset{*}{u}_{1,2} + W_2'\overset{*}{\Lambda}_2 + \dot{W}_2\epsilon_2 + W_{22} = 0$$
$$\Delta_1\overset{*}{u}_{1,2} + \dot{\Delta}\epsilon_2 + \Delta_2 = 0$$

which upon evaluation at B (where $W_{22} = 0$) results in

$$\epsilon_2 = 0, \quad (\text{assuming } \dot{W}_2 \neq 0)$$
$$\overset{*}{u}_{1,2} = 0 \quad \text{and} \quad \overset{*}{\Lambda}_2 = 0 \tag{1.3.56}$$

A second perturbation yields

$$W_{11}\overset{*}{u}_{1,22} + W_1'\overset{*}{\Lambda}_{22} + W_{221} = 0$$
$$\epsilon_{22} = 0 \tag{1.3.57}$$

and

$$\overset{*}{u}_{1,22} = -\left(\frac{W_{2222}}{W_{221}} - 2\frac{W_{221}}{W_{11}}\right)$$

which can be solved for $\overset{*}{\Lambda}_{22}$ to give

$$\overset{*}{\Lambda}_{22} = \frac{1}{W_1'}\left(\frac{W_{11}W_{2222}}{W_{221}} - 3W_{221}\right) \tag{1.3.58}$$

It is noted in passing that

$$\overset{*}{\Lambda}_{22} = 3\Lambda_{22}$$

Finally, a third perturbation of (1.3.53) results in

$$\epsilon_{222} = \frac{2}{\dot{W}_2 W_{11}}[W_{11}W_{2222} - 3(W_{221})^2] \tag{1.3.59}$$

Asymptotic equations of the *critical line* are now readily constructed as

$$\overset{*}{u}_1 = u_1^B - \frac{1}{2}\left(\frac{W_{2222}}{W_{221}} - 2\frac{W_{221}}{W_{11}}\right)\overset{*}{y}_2^2$$
$$\overset{*}{\Lambda} = \Lambda^B + \frac{1}{2W_1'}\left(\frac{W_{11}W_{2222}}{W_{221}} - 3W_{221}\right)\overset{*}{y}_2^2 \tag{1.3.60}$$

and

$$\epsilon = \frac{1}{3\dot{W}_2 W_{11}}[W_{11}W_{2222} - 3(W_{221})^2]\overset{*}{y}_2^3$$

1.3 Symmetric systems: simple critical points

which results in the important critical load-imperfection relationship

$$\overset{*}{\Lambda} = \Lambda^B + \tfrac{3}{2}\Lambda_{22}^{\frac{1}{3}}\left(\frac{\dot{W}_2 W_{11}}{W'_1 W_{221}}\epsilon\right)^{\frac{2}{3}} \tag{1.3.61}$$

These results will be used directly in the analysis of a shallow arch in Section 1.4.5.

Problem. In the above analysis, asymptotic equations of the critical line (1.3.60) have been obtained directly, by-passing the equilibrium paths of the imperfect system. Formulate the problem concerning the derivation of the equations of such paths through a two-parameter perturbation technique and construct the paths asymptotically. Introduce, then, the critical condition, $\Lambda = 0$, to recover the critical line (1.3.60) by solving the latter with the equilibrium equations concurrently.

1.4

Examples: simple critical points

1.4.1 General remarks

In Chapters 1.1, 1.2 and 1.3, a general nonlinear theory concerning the equilibrium and stability of discrete conservative systems is presented. Critical points arising throughout this theory are assumed to be *simple*, and it is evident that four such points, namely, *limit points, asymmetric, stable-symmetric* and *unstable-symmetric bifurcation points* are of paramount importance in the theory of elastic stability. The applications in this chapter are, therefore, designed to illustrate various concepts of the presented theory with particular reference to these critical points.

Rigid link models, lending themselves to a comparatively simple treatment, are often instrumental in acquiring a deeper insight into general theories, and will be used here for this purpose. On the other hand, among engineering structures, the shallow arch exhibits typical nonlinear buckling and post-buckling characteristics which are often useful for illustrating many nonlinear concepts of the general theory. Shallow circular arches under various loading conditions will, therefore, be analyzed here as well as in Part 2 with a view to demonstrating various aspects of the general theory.

1.4.2 Limit points: one-degree-of-freedom model

Consider the simple rigid link model (first used by von Mises) shown in Figure 1.16. The model consists of two equal pin-jointed straight members each rigid in bending but permitting linearly elastic axial deformations (with a spring constant k). The loading on the frame consists of a vertical concentrated load, Λ, at the centre.

It is assumed that the frame is sufficiently shallow and deflects symmetrically, so all configurations can be described by a single generalized coordinate, the angle Q_1.

1.4 Examples: simple critical points

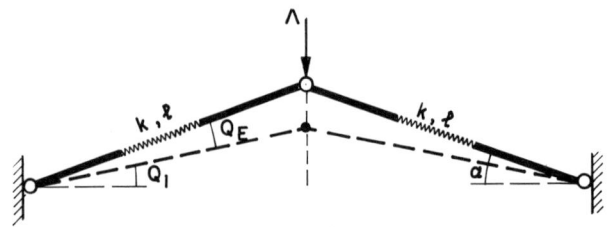

Figure 1.16

From geometry, one has

$$\Delta l = l\left(1 - \frac{\cos \alpha}{\cos Q_1}\right)$$
$$E = l(\sin \alpha - \cos \alpha \operatorname{tg} Q_1) \tag{1.4.1}$$

where E is the deflection corresponding to Λ.
The potential energy of the system is

$$V(Q_1, \Lambda) = U(Q_1) - \Lambda E(Q_1) = kl^2\left(1 - \frac{\cos \alpha}{\cos Q_1}\right)^2$$
$$- \Lambda l(\sin \alpha - \cos \alpha \operatorname{tg} Q_1) \tag{1.4.2}$$

The equilibrium equation $\partial V/\partial Q_1 = 0$ then, yields

$$\Lambda = 2kl \operatorname{tg} Q_1(\cos Q_1 - \cos \alpha) \tag{1.4.3}$$

For small Q_1 and α (shallow frame) one obtains

$$\Lambda = klQ_1(\alpha^2 - Q_1^2) \tag{1.4.4}$$

which describes the path shown in Figure 1.17.

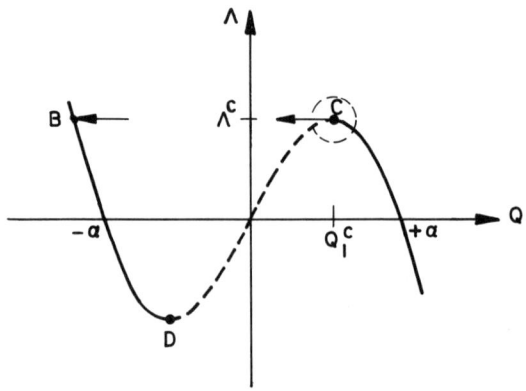

Figure 1.17

1.4.2 Limit points: one-degree-of-freedom model

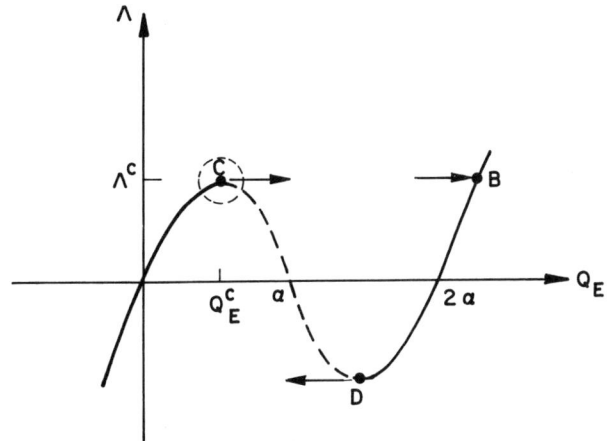

Figure 1.18

The analysis is facilitated by the choice of Q_1 as the generalized coordinate; however, $Q_1 = \Lambda = 0$ does not, in this case, represent the unloaded zero state, and one might like to shift to zero-state by introducing $Q_E = \alpha - Q_1$ into (1.4.4). The equilibrium path in $\Lambda - Q_E$ plane, then, takes the expected typical form as shown in Figure 1.18 and described by

$$\frac{\Lambda}{kl} = Q_E(\alpha - Q_E)(2\alpha - Q_E) \qquad (1.4.5)$$

Critical loads can be obtained by locating the extremum points on the path as

$$\frac{\Lambda^c}{kl} = \pm \frac{2\sqrt{3}}{9} \alpha^3 \qquad (1.4.6)$$

by making use of

$$Q_1^c = \pm \frac{\sqrt{3}}{3} \alpha \quad \text{or} \quad Q_E^c = \alpha\left(1 \pm \frac{\sqrt{3}}{3}\right)$$

It is easy to show that the third derivative of the energy function at C (and at D) does not vanish, and the critical point C (and D) is, therefore, unstable. When the loading parameter Λ is increased gradually from zero, the frame deflects symmetrically, and an initially stable but nonlinear path is traced in the $\Lambda - Q_E$ load-deflection space (Figure 1.18) until the limit point C is reached. The configuration of the frame at C is prescribed by $Q_E^c = \alpha\left(1 - \frac{\sqrt{3}}{3}\right)$, and at this unstable extremum point a slight perturbation makes the frame snap to the configuration of point B which is stable.

1.4 Examples: simple critical points

At B, the frame has a reversed curvature and the two identical members are in tension. An analogous behaviour is observed if the frame is unloaded gradually from $\Lambda^c = \dfrac{2\sqrt{3}}{9}\alpha^3$; the jump, however, now occurs at D. It is worth noting that $Q_E = \alpha$ lies on an unstable portion of the path and will snap-through to the configuration of point $Q_E = 2\alpha$ or $Q_E = 0$ depending on the disturbance.

Problem 1. Show analytically that the point $Q_1 = 0$ (or $Q_E = \alpha$) is unstable (a) by first assuming that Q_1 and α are small, (b) and then, without making the above assumption. Also determine the critical load without assuming that Q_1 and α are small.

Problem 2. Introduce an initial small imperfection into the geometry of the frame by assuming that initial angle $\alpha_0 = \alpha + \epsilon$, and derive the first order critical $\overset{*}{\lambda} - \epsilon$ relationship. Show that on a plot of $\overset{*}{\lambda}$ versus ϵ, the curve has a finite slope at $\epsilon = 0$. Compare this problem with example 1 in Section 2.2.8. (Can you treat Λ' as an imperfection parameter?)

1.4.3 Asymmetric points of bifurcation: one-degree-of-freedom model

Another simple structural model concerning the asymmetric points of bifurcation will now be analyzed (see also [41]).

The model consists of a rigid rod of length l hinged at the bottom and supported by a linear inclined spring (with constant k) at the top. A vertical concentrated load of magnitude Λ is applied eccentrically, and the eccentricity is assumed to be ϵl (Figure 1.19). Obviously, the system has

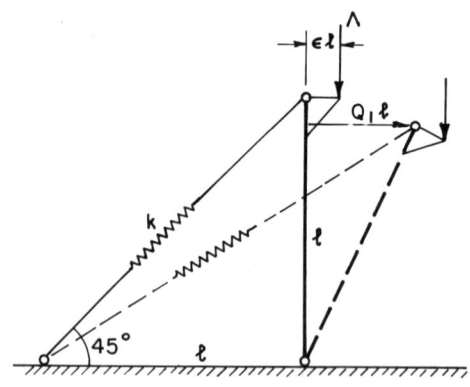

Figure 1.19

1.4.3 Asymmetric points of bifurcation: one-degree-of-freedom model

only one degree of freedom, and choosing the horizontal displacement of the top of the rod as $Q_1 l$, all configurations of the system may be specified with the coordinate Q_1.

The total potential energy of the imperfect system can be expressed as

$$V(Q_1, \Lambda, \epsilon) = kl^2[(1+Q_1)^{\frac{1}{2}} - 1]^2 - \Lambda l[1 - (1-Q_1^2)^{\frac{1}{2}} + \epsilon Q_1] \tag{1.4.7}$$

which, by virtue of the Binomial Theorem, takes the form

$$V = kl^2(\tfrac{1}{2}Q_1 - \tfrac{1}{8}Q_1^2 + \cdots)^2 - \Lambda l(\tfrac{1}{2}Q_1^2 + Q_1\epsilon + \cdots) \tag{1.4.8}$$

Evidently, $Q_1 = 0$ describes the fundamental path of the perfect system, and the only operation required for transforming the function V into the function S of Section 1.1.6 consists of setting $Q_1 = u_1$ in (1.4.7) or (1.4.8). Then, $S(u_1, \Lambda, \epsilon)$ leads to the results of the general theory for this special case. Thus, differentiating (1.4.8) with respect to u_1 one obtains

$$S_{11}(0, \Lambda, 0) = \tfrac{1}{2}kl^2 - \Lambda l$$

yielding

$$\Lambda^c = \tfrac{1}{2}kl \tag{1.4.9}$$

Similarly,

$$\dot{S}_1(0, \Lambda^c, 0) = -\tfrac{1}{2}kl^2$$
$$S_{111}(0, \Lambda^c, 0) = -\tfrac{3}{4}kl^2 \tag{1.4.10}$$

and

$$S'_{11}(0, \Lambda^c, 0) = -l$$

The slope of the post-buckling path of the perfect system is

$$\Lambda_1 = -\frac{S_{111}}{2S'_{11}} = -\tfrac{3}{8}kl \tag{1.4.11}$$

The critical load-imperfection relationship in the vicinity of Λ^c can readily be obtained by using the theoretical results of Section 1.2.3 as

$$\overset{*}{\lambda} = \pm\tfrac{1}{2}kl(3\epsilon)^{\frac{1}{2}} \quad \text{from (1.2.16)}$$

or $\tag{1.4.12}$

$$\overset{*}{\Lambda} = \tfrac{1}{2}kl[1 \pm (3\epsilon)^{\frac{1}{2}}] \quad \text{from (1.2.27)}$$

The equilibrium paths of the imperfect system in the vicinity of the bifurcation point can be obtained from the equilibrium surface (1.2.9) which in this problem takes the form

$$\frac{kl}{2}\epsilon + \tfrac{3}{8}klu_1^2 + u_1\lambda = 0 \tag{1.4.13}$$

1.4 Examples: simple critical points

For a specified value of ϵ, one obtains the equilibrium paths of the corresponding imperfect system. It is worth noting that the critical load-imperfection relationship (1.4.12) can also be derived from (1.4.13) by recalling that critical points on this surface represent the extrema. Thus, differentiating (1.4.13) with respect to u_1 and setting equal to zero, one has

$$\tfrac{3}{4}kl\overset{*}{u}_1 + \overset{*}{\lambda} = 0 \qquad (1.4.14)$$

which when substituted for u_1 into (1.4.13) results in the first relation in (1.4.12).

Setting $\epsilon = 0$ in (1.4.13) yields the fundamental path $u_1 = 0$ and the post-buckling path

$$\tfrac{3}{8}klu_1 + \lambda = 0 \qquad (1.4.15)$$

which shows on comparison with (1.4.14) that

$$\overset{*}{\lambda}_1 = 2\lambda_1 \quad \text{or equivalently} \quad \overset{*}{\Lambda}_1 = 2\Lambda_1$$

as predicted in the theory (Section 1.2.3). The slope of the post-buckling path is negative and for $\epsilon > 0$, (1.4.12) yields the critical limit points of the imperfect system as depicted in Figure 1.20. It is further observed that (1.4.12) with the negative sign gives the actual critical points arising in a natural sequence of loading from zero while the same equation with the positive sign yields the extrema on the complementary paths. For $\epsilon < 0$ there are, of course, no critical points.

Problem 1. Discuss the stability of the bifurcation point, and the stability distribution on the fundamental and post-buckling paths of the perfect system by following the general theory presented in Section 1.1.7.

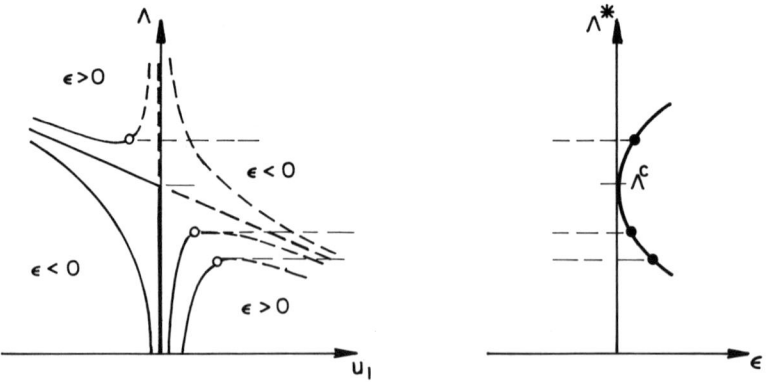

Figure 1.20

1.4.4 Stable-symmetric points of bifurcation

Problem 2. Generate another family of imperfect systems by assuming that in addition to the eccentricity in loading, the spring is initially too long by $\sqrt{2}\,dl\epsilon$ where d is a constant. Express the total potential energy of this family and derive the asymptotic equation of the equilibrium surface in the vicinity of the bifurcation point. Show that the critical load-imperfection relationship is given by $\Lambda = \frac{1}{2}kl\{1 \pm 3(1+2d)\epsilon]^{\frac{1}{2}}\}$.

1.4.4 Stable-symmetric points of bifurcation: two-degree-of-freedom model

To illustrate various features of stable-symmetric points of bifurcation consider the two-degree-of-freedom rigid link model (Figure 1.21)

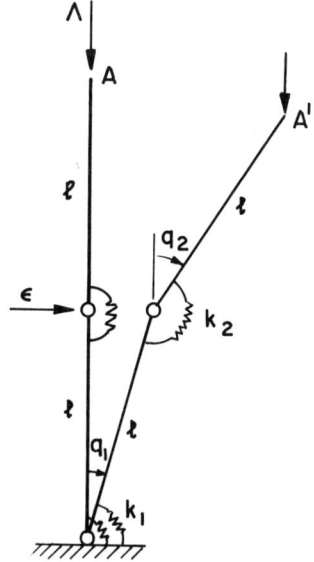

Figure 1.21

consisting of two pin-jointed straight members. Each of the two joints has a linear torsional spring of stiffnesses $k_1 = 2k$ and $k_2 = k$. A concentrated vertical load of magnitude Λ acts at A as shown, and a small horizontal force ϵ acting at B renders the system imperfect.

The total potential energy of the system is

$$V = kq_1^2 + \tfrac{1}{2}k(q_2 - q_1)^2 - \Lambda l(2 - \cos q_1 - \cos q_2)$$
$$- \epsilon l \sin q_1 \qquad (1.4.16)$$

67

1.4 Examples: simple critical points

which, after using trigonometric series, takes the form

$$V = kq_1^2 + \tfrac{1}{2}k(q_2-q_1)^2$$
$$- \Lambda l[\tfrac{1}{2}(q_1^2+q_2^2) - \tfrac{1}{24}(q_1^4+q_2^4) + \cdots]$$
$$- \epsilon l\left(q_1 + \frac{1}{3!}q_1^3 + \cdots\right) \qquad (1.4.17)$$

It is obvious from the physical system that the fundamental path of the perfect system is defined by $Q_i = 0$ and the Q_i are, therefore, replaced by the q_i in the energy expression. In order to transform the function $V(q_i, \Lambda, \epsilon)$ into the function $S(u_i, \Lambda, \epsilon)$, introduce the orthogonal transformation

$$q_i = \alpha_{ij} u_j, \qquad \alpha_{ij}\alpha_{jk} = \delta_{ik} \qquad (1.4.18)$$

where

$$\alpha_{ij} = \begin{bmatrix} 0.3827 & 0.9239 \\ 0.9239 & -0.3827 \end{bmatrix}$$

to diagonalize the quadratic form of V in the q_i.

Substituting (1.4.18) into (1.4.17), then, yields

$$S(u_i, \Lambda, \epsilon) = \tfrac{1}{2}k(0.59u_1^2 + 3.41u_2^2) - \Lambda l\{\tfrac{1}{2}(u_1^2+u_2^2)$$
$$- \tfrac{1}{24}[(0.38u_1 + 0.92u_2)^4 + (0.92u_1 - 0.38u_2)^4]\}$$
$$- \epsilon l\left[0.38u_1 + 0.92u_2 + \frac{1}{3!}(0.38u_1 + 0.92u_2)^3 + \cdots\right] \qquad (1.4.19)$$

Differentiating this function as required one obtains the essential derivatives

$$\begin{aligned}S_{11}(0, \Lambda, 0) &= 0 \cdot 59k - l\Lambda \\ S_{22}(0, \Lambda, 0) &= 3 \cdot 41k - l\Lambda\end{aligned} \qquad (1.4.20)$$

yielding the critical load

$$\Lambda^c = 0 \cdot 59k/l \qquad (1.4.21)$$

and

$$S_{22}(0, \Lambda^c, 0) = 2 \cdot 82k \qquad (1.4.22)$$

Similarly,

$$\begin{aligned} S_{111} &= 0, & S_{1111} &= 0 \cdot 44k, & S_{211} &= 0 \\ S'_{11} &= -l, & \dot{S}_1 &= -0 \cdot 38l, & \dot{S}_2 &= -0 \cdot 92l \end{aligned} \qquad (1.4.23)$$

1.4.4 Stable-symmetric points of bifurcation

Using now (1.2.11) and (1.2.12) from the general theory, one constructs the asymptotic equations of the equilibrium surface which take the form

$$-0{\cdot}381\epsilon - lu_1\lambda + \frac{0{\cdot}44}{3!}ku_1^3 = 0$$
$$2{\cdot}82ku_2 + 2{\cdot}42u_1\lambda = 0$$
(1.4.24)

The initial curvature of the post-buckling path here is given by

$$\Lambda_{11} = -\frac{S_{1111}}{3S'_{11}} = 0{\cdot}15k/l \tag{1.4.25}$$

Obviously, the symmetric bifurcation point is a stable one, and the neighbouring equilibrium paths of the imperfect system are in the form illustrated in Figure 1.14. If the slightly imperfect system is loaded gradually, the equilibrium paths of the imperfect system raise continuously, no instability being exhibited. However, for appropriate combinations of $\Lambda(\Lambda > \Lambda^c)$ and ϵ, limit points may arise on the complementary paths, and the critical load-imperfection relationship is readily derived on the basis of the general theory as

$$\overset{*}{\Lambda} = 0{\cdot}59k/l + \tfrac{1}{2}(0{\cdot}44k)^{\frac{1}{3}}\frac{(3\times 0{\cdot}38l\epsilon)^{\frac{2}{3}}}{l}$$

by using (1.2.28).

Finally, it is observed that the total potential energy function of the perfect system ($\epsilon = 0$) is symmetric in the generalized coordinates as defined in Section 1.3.2, explaining why $S_{211} = S_{111} = 0$. The next example is also concerned with a symmetric structure.

Problem 1. Examine the stability of the bifurcation point associated with the perfect system and show that this point is stable through the formal procedure described in Section 1.1.4. Also examine the stability distribution on the fundamental and post-buckling paths in the vicinity of the bifurcation point and show that the post-buckling path is totally stable.

Problem 2. An interesting model with two degrees of freedom is discussed by Thompson. The model is a rigid prismatic body of mass m floating in an ocean of density ρ. Assuming that the right-angled prism is of unit length, taking the generalized coordinates as shown in Figure 1.22, and the height of the centre of gravity above the apex as the loading parameter generate a family of imperfect systems by off-setting the centre of gravity a distance ϵ. Analyze the imperfect system fully and show that asymptotic variation of the critical load with the imperfection parameter

1.4 Examples: simple critical points

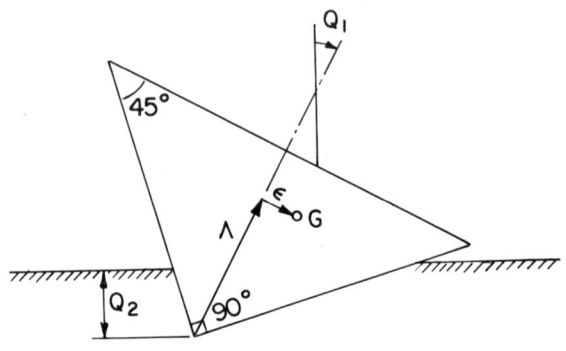

Figure 1.22

is given by

$$\overset{*}{\Lambda} = \frac{4}{3}\left(\frac{m}{\rho}\right)^{\frac{1}{2}}\left[1 + 1\cdot 24\left(\frac{\rho}{m}\right)^{\frac{1}{3}}\epsilon^{\frac{2}{3}}\right].$$

What is the bifurcation load of the perfect system? Construct the asymptotic equation of the equilibrium surface in the vicinity of the bifurcation point and obtain the above $\overset{*}{\Lambda} - \epsilon$ relation from this equation.

1.4.5 Shallow arch: unstable-symmetric points of bifurcation and limit points

The general theory shows that in the vicinity of an unstable-symmetric point of bifurcation the critical loads of the associated imperfect systems may be reduced considerably for both positive and negative values of the imperfection parameter. This fourth example, namely the shallow circular arch, is capable of illustrating this phenomenon as well as various other aspects of the general theory.

Many authors [36, 65, 66] have analyzed shallow arches with various boundary conditions, and under certain external loads. An exact analysis can be found in [65].

The buckling and post-buckling behaviour of a simply-supported circular arch under the *combined* effect of more than one independent load was discussed in [43, 44, 48, 50] and will be treated in Part 2.

Here emphasis is on imperfections, and a family of imperfect systems is generated by off-setting the load by a small angle ϵ with the vertical [36] as shown in Figure 1.23. The circular arch considered is of radius R with a central angle $2\theta_0$ and it is simply supported. The discussion is limited to a shallow arch whose rise at the crown is much smaller than R. It is further assumed that the arch has a constant cross

1.4.5 Shallow arch: unstable-symmetric points of bifurcation

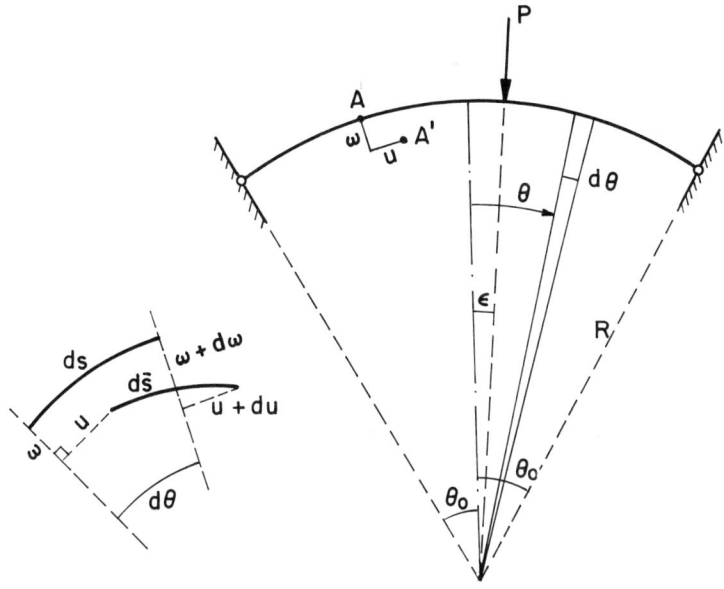

Figure 1.23

section of area A, of thickness h and width unity (i.e., $A = h \times 1$). For a shallow arch

$$u \ll R$$
$$\omega \ll R \tag{1.4.26}$$

where u and ω denote tangential and radial displacements respectively. Since $(\omega_\theta/R)^2 \ll 1$, where $\omega_\theta \equiv d\omega/d\theta$, the change in curvature, χ, and axial strain, e, are given approximately by

$$\chi = \frac{1}{R^2} \omega_{\theta\theta}$$

and (1.4.27)

$$e = \frac{1}{R}(u_\theta - \omega) + \frac{1}{2R^2} \omega_\theta^2$$

respectively, where the subscript θ denotes partial differentiation with respect to θ.

The latter expression is derived readily by recognizing that

$$e = \frac{d\bar{s} - ds}{ds}$$

1.4 Examples: simple critical points

where $d\bar{s}$ is the deformed length of an arch element $ds = R\,d\theta$, and substituting for $d\bar{s}$ the approximate value

$$d\bar{s} = \frac{1}{R}[(R-\omega+u_\theta)^2 + \omega_\theta^2]^{\frac{1}{2}}\,ds$$

$$\approx \left(1 + \frac{u_\theta - \omega}{R} + \frac{\omega_\theta^2}{2R^2}\right) ds$$

which is evident from Figure 1.23.

The strain energy is the sum of the energies due to axial deformations and bending, and can be written as

$$U = \frac{EAR}{2}\int_{-\theta_0}^{+\theta_0} e^2\,d\theta + \frac{EIR}{2}\int_{-\theta_0}^{+\theta_0} \chi^2\,d\theta$$

where E is the Young's modulus of elasticity, I is the moment of inertia and θ_0 half the central angle of the arch.

The potential energy of the external load is given by

$$-P\omega|_{\theta=\epsilon}.$$

Before proceeding to construct the total potential energy function, it is observed that only the strain energy due to axial deformations contains the displacement variable u, and it can be eliminated from the energy expression altogether on the basis of the assumptions (1.4.26).

In fact, e can be treated as constant along the axis, and the integration of the expression for axial strain in (1.4.27) between $-\theta_0$ and $+\theta_0$, therefore, yields

$$e\theta\Big]_{-\theta_0}^{+\theta_0} = \frac{1}{R}u\Big]_{-\theta_0}^{+\theta_0} - \frac{1}{R}\int_{-\theta_0}^{+\theta_0}\omega\,d\theta + \frac{1}{2R^2}\int_{-\theta_0}^{+\theta_0}\omega_\theta^2\,d\theta$$

Using the boundary conditions $u = 0$ at $\theta = -\theta_0$ and $\theta = +\theta_0$ one has

$$e = \frac{1}{2\theta_0}\int_{-\theta_0}^{+\theta_0}\left(-\frac{\omega}{R} + \frac{\omega_\theta^2}{2R^2}\right)d\theta$$

With the use of this expression the total potential energy function can be written as

$$V = \frac{1}{4\theta_0^2}\left[\int_{-\theta_0}^{+\theta_0}(-\bar{\omega} + \tfrac{1}{2}\bar{\omega}_\theta^2)\,d\theta\right]^2 + \frac{I}{2R^2A\theta_0}\int_{-\theta_0}^{+\theta_0}\bar{\omega}_{\theta\theta}^2\,d\theta - \Lambda\bar{\omega}\Big|_{\theta=\epsilon} \quad (1.4.28)$$

which is nondimensionalized through dividing by $EAR\theta_0$, and where $\bar{\omega} = \omega/R$, $\Lambda = P/EA\theta_0$.

This energy function is now independent of u, thus allowing for a

1.4.5 Shallow arch: unstable-symmetric points of bifurcation

Rayleigh–Ritz type of analysis based on an assumed displacement function $\bar{\omega}$ only. One can, for example, assume that $\bar{\omega}$ is represented by the series

$$\bar{\omega} = \sum_{i=1,3,5,\ldots} Q_i \cos\frac{i\pi\theta}{2\theta_0} + \sum_{\alpha=2,4,6,\ldots} z_\alpha \sin\frac{\alpha\pi\theta}{2\theta_0}$$

which is an admissible function and satisfies the dynamic as well as geometric boundary conditions.

Substituting for $\bar{\omega}$ in the energy function (1.4.28) and evaluating the integrals yield an energy expression similar to (1.3.4) for the perfect system ($\epsilon = 0$), indicating that the system under consideration falls within the scope of the general theory presented in Chapter 1.3 in its idealized form. For the purpose of illustration here, however, it will suffice to take the first term from each of the sine and cosine series and assume a deflection pattern in the form

$$\bar{\omega} = u_1 \cos\frac{\pi\theta}{2\theta_0} + y_2 \sin\frac{\pi\theta}{\theta_0} \qquad (1.4.29)$$

in which Q_1 and z_2 are directly replaced by u_1 and y_2 respectively, the transformations (1.3.8) not being required.

Substituting now for $\bar{\omega}$ into (1.4.28) and observing that

$$\bar{\omega}|_{\theta=\epsilon} \approx u_1 + 2c\epsilon y_2$$

since ϵ is very small, one obtains

$$W(u_1, y_2, \Lambda, \epsilon) = \frac{1}{4}\left(\frac{1}{2}c^2 u_1^2 + 2c^2 y_2^2 - \frac{4}{\pi}u_1\right)^2 + d(\tfrac{1}{2}c^4 u_1^2 + 8c^4 y_2^2)$$
$$- \Lambda u_1 - 2c\Lambda\epsilon y_2 \qquad (1.4.30)$$

in which

$$c = \frac{\pi}{2\theta_0}, \qquad d = \frac{I}{R^2 A}.$$

Evidently, for $\epsilon = 0$, the perfect system falls within the scope of the formulation in Section 1.3.2 and the results of that section as well as the subsequent sections in 1.3 are directly applicable.

A significant distinction of this problem from the preceding examples treated in Sections 1.4.3 and 1.4.4 is that the fundamental path associated with (1.4.30) is not identified by $u_1 = y_2 = 0$; on the contrary the path is highly nonlinear and exhibits limit points in the $u_1 - \Lambda$ plane. Consequently, only y_2 may be considered as sliding along the fundamental path, and the function W does not have the properties of the function S employed in earlier examples.

1.4 Examples: simple critical points

Following the general theory of Section 1.3.3, ordered estimates of the fundamental path can readily be obtained. To this end, evaluate first the energy derivatives at the origin

$$W_{11} = \frac{8}{\pi^2} + dc^4, \qquad W_{22} = 16dc^4,$$

$$W_{111} = -\frac{6c^2}{\pi}, \qquad W_{122} = -\frac{8c^2}{\pi},$$

$$W_{1111} = \tfrac{3}{2}c^4, \qquad W_{1122} = 2c^4,$$

$$W'_1 = -1, \qquad W'_2 = 0.$$

The path derivatives (1.3.14), (1.3.15) are in the form

$$\Lambda_1 = \frac{8}{\pi^2} + dc^4$$

$$\Lambda_{11} = -\frac{6c^2}{\pi}$$

and

$$\Lambda_{111} = \tfrac{3}{2}c^4$$

which can be used in conjunction with (1.3.16) to obtain the first, second and third order estimates of the fundamental path as

$$\Lambda = \left(\frac{8}{\pi^2} + dc^4\right)u_1$$

$$\Lambda = \left(\frac{8}{\pi^2} + dc^4\right)u_1 - \frac{3c^2}{\pi}u_1^2 \qquad (1.4.31)$$

and

$$\Lambda = \left(\frac{8}{\pi^2} + dc^4\right)u_1 - \frac{3c^2}{\pi}u_1^2 + \tfrac{1}{4}c^4 u_1^3$$

respectively. This last equation could, of course, be obtained by differentiating the energy function (1.4.30) with respect to u_1 and y_2, and solving the resulting equilibrium equations simultaneously. For a two-degree-of-freedom system this procedure presents no difficulties; in the case of several degrees of freedom, however, it becomes unmanageable while the perturbation approach preserves its essential systematic character.

For estimates of stability limits, first use (1.3.23), (1.3.24) and (1.3.25) to determine

$$\Lambda_0^u = \frac{8}{\pi^2} + dc^4$$

$$\Lambda_1^u = -\frac{6c^2}{\pi}$$

1.4.5 Shallow arch: unstable-symmetric points of bifurcation

and

$$A_{11}^u = \tfrac{3}{2}c^4$$

which yield the ordered estimates

$$\frac{8}{\pi^2} + dc^4 - \frac{6c^2}{\pi} u_1 = 0$$

and

$$\frac{8}{\pi^2} + dc^4 - \frac{6c^2}{\pi} u_1 + \frac{3}{4} c^4 u_1^2 = 0$$

The last equation gives two limit points on the fundamental path

$$u_1^L = \frac{4}{c^2}\left[\frac{1}{\pi} \pm \frac{1}{3}\left(\frac{3}{\pi^2} - \frac{3c^4 d}{4}\right)^{\frac{1}{2}}\right] \tag{1.4.32}$$

Here u_1^L with the negative sign represents the stability limit.

Next, use (1.3.29) to obtain

$$A_0^y = 16dc^4$$

$$A_1^y = -8\frac{c^2}{\pi}$$

$$A_{11}^y = 2c^4$$

which, in turn, are used to construct the ordered estimates of bifurcation points

$$16dc^4 - 8\frac{c^2}{\pi} u_1 = 0$$

and

$$16dc^4 - 8\frac{c^2}{\pi} u_1 + c^4 u_1^2 = 0.$$

The last equation yields two bifurcation points on the fundamental path

$$u_1^B = \frac{4}{c^2}\left[\frac{1}{\pi} \pm \left(\frac{1}{\pi^2} - c^4 d\right)^{\frac{1}{2}}\right] \tag{1.4.33}$$

Corresponding limit and bifurcation loads can be obtained by substituting for u_1^L and u_1^B (with minus sign) into the third order equilibrium equation (1.4.31).

The bifurcation load, for example, can thus be obtained as

$$\Lambda^B = \frac{4c^2 d}{\pi} + 12c^2 d\left(\frac{1}{\pi^2} - c^4 d\right)^{\frac{1}{2}} \tag{1.4.34}$$

1.4 Examples: simple critical points

Since the fundamental path is represented by a single-valued function of u_1, and monotonically increasing u_1 describes the path, one condition that bifurcation occurs before the limit point is

$$u_1^B < u_1^L$$

where u_1^B and u_1^L are the smaller critical values (i.e., with negative signs in (1.4.32) and (1.4.33)).

This condition is satisfied if

$$\left(\frac{1}{\pi^2} - c^4 d\right)^{\frac{1}{2}} > \frac{1}{3}\left(\frac{3}{\pi^2} - \frac{3c^4 d}{4}\right)^{\frac{1}{2}}$$

which yields

$$\frac{8}{11\pi^2} > c^4 d \qquad (1.4.35)$$

(1.4.35) is a sufficient condition as well as a necessary one since it also ensures real u_1^B.

Post-buckling behaviour

Suppose now (1.4.35) holds; the next question is, then, concerned with the post-buckling path through the bifurcation point and ultimately with the behaviour of the associated imperfect system. Such an analysis can readily be performed since the results of Section 1.3.5 are now directly applicable. For this purpose, evaluate first the energy derivatives at the bifurcation point B (Figure 1.24),

$$W_{11} = \frac{8}{\pi^2} - 11 dc^4$$

$$W_{12} = W_{22} = 0$$

$$W_1' = -1, \qquad \dot{W}_1 = 0$$

$$W_{122} = -8c^2 \left(\frac{1}{\pi^2} - c^4 d\right)^{\frac{1}{2}} \qquad (1.4.36)$$

$$W_{2222} = 24 c^4$$

$$\dot{W}_2 = -8c^3 d \left[\frac{1}{\pi} + 3\left(\frac{1}{\pi^2} - c^4 d\right)^{\frac{1}{2}}\right]$$

First order equations of the post-buckling path are given by (1.3.49)

1.4.5 Shallow arch: unstable-symmetric points of bifurcation

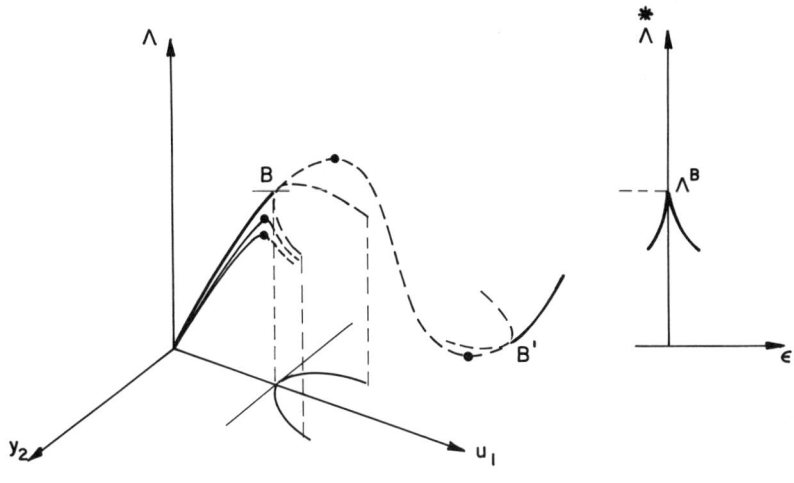

Figure 1.24

and (1.3.50) which upon substitutions from (1.4.36) take the form

$$u_1 = u_1^B + \frac{c^2}{2}\left(\frac{1}{\pi^2} - c^4 d\right)^{-\frac{1}{2}} y_2^2$$
$$\Lambda = \Lambda^B - \tfrac{3}{2} c^6 d\left(\frac{1}{\pi^2} - c^4 d\right)^{-\frac{1}{2}} y_2^2 \qquad (1.4.37)$$

in which u_1^B and Λ^B are given by (1.4.33) and (1.4.34) respectively (with negative signs).

Similarly, the asymptotic equation of the *critical line* of the imperfect system can be obtained by merely substituting for appropriate derivatives into (1.3.60) from (1.4.36). The critical load-imperfection relationship (1.3.61) takes the form

$$\overset{*}{\Lambda} = \Lambda^B + \tfrac{3}{2}\Lambda_{22}^{\frac{1}{3}} \left\{ -\frac{cd\left[\frac{1}{\pi} + 3\left(\frac{1}{\pi^2} - c^4 d\right)^{\frac{1}{2}}\right]\left(\frac{8}{\pi^2} - 11 c^4 d\right)}{\left(\frac{1}{\pi^2} - c^4 d\right)^{\frac{1}{2}}} \epsilon \right\}^{\frac{2}{3}}$$

where Λ_{22} is the curvature of the post-buckling path (1.4.37). This relation and the equilibrium paths of the imperfect system are shown in Figure 1.24 which indicates the imperfection sensitivity of the system.

Problem 1. Perform a post-buckling and imperfection analysis in the vicinity of B' (Figure 1.24) and sketch the equilibrium paths of the imperfect system in that neighbourhood. Show that the projection of the

1.4 Examples: simple critical points

post-buckling path onto $u_1 - \Lambda$ plane is in the form of a straight line, and combine all the results to predict and sketch the overall post-buckling paths and the paths of the imperfect system.

Problem 2. Introduce a small imperfection in another form and perform a similar buckling, post-buckling and imperfection analysis.

1.5

Coincident critical points

1.5.1 General remarks

So far the theory and applications have been concerned solely with simple (discrete) critical points and related stability problems in which the primary critical point (or the critical point of interest) lies distinctly apart from other critical points on the fundamental path, and there is a well-defined buckling mode corresponding to the critical load. Normally, the fundamental path is intersected by a unique post-buckling path at a simple critical point of bifurcation (limit points are excluded from the discussion in this chapter). Although simple critical points arise commonly in the theory of elastic stability, a number of important problems are associated with *coincident* critical points (or nearly so) at which so-called *simultaneous buckling* occurs. In the load-deflection space a coincident critical point on the fundamental path often manifests itself as a *multiple path generator* in mixed modes, resulting in a complex behaviour.

Axially compressed columns on elastic foundations, rectangular plates and cylindrical shells, for example, can exhibit such a *compound branching* phenomenon under certain circumstances. Koiter [12, 67] discussed the effect of axisymmetric imperfections on the buckling of cylindrical shells under axial compression, and showed that compound branching is responsible for greater sensitivity to imperfections. Stein [68] and Augusti [69] presented rigid link models with two degrees of freedom exhibiting coincident critical points. Chilver [70] and Supple [71, 72] explored various possible post-buckling path configurations in the vicinity of *nearly coincident* (as well as coincident) critical points, the latter author focussing his attention on symmetric systems. These results were generalized to N-degree-of-freedom systems by Chilver and Johns [73, 74, 75] who considered asymmetric, partly symmetric and completely symmetric cases. Sewell [76] discussed a systematic perturbation method leading to the choice of an appropriate perturbation parameter for branching analysis. Thompson and Hunt [77] proposed a method for numerical analysis.

1.5 Coincident critical points

In Ref. [13], Koiter presented a full rigorous discussion concerning the stability of coincident critical points for continuous systems. The same problem for discrete systems was tackled by Sewell [78] via the notion of 'order of contact'. A multiple-parameter perturbation technique was later introduced as an alternative method of analysis [52, 53] which is in line with compound branching conditions, and will now be presented.

1.5.2 Stability of coincident critical points

Suppose the potential energy $H(u_i, \lambda)$ given by (1.1.10) is referred to a coincident critical point, F, at which several stability coefficients vanish simultaneously, resulting in *simultaneous buckling* or equivalently *compound branching*. In analogy with Section 1.1.4, the convexity of the energy surface at F is to be examined. To this end, assume that

$$H_{\alpha\alpha} = 0 \quad \text{where} \quad \alpha = 1, 2, \ldots, n; \quad n < N$$
and (1.5.1)
$$H_{ss} > 0 \quad \text{where} \quad s = n+1, \ldots, N;$$

implying the notations u_α and u_s for critical and non-critical coordinates respectively.

Considering the path $u_i = u_i(\eta)$ given by (1.1.33) again, the change in energy along this path is expressed as

$$h(\eta) = H[u_i(\eta), 0] - H(0, 0) \tag{1.5.2}$$

Then,

$$\left.\frac{dh}{d\eta}\right|_F = H_i u_{i,\eta} = 0$$

since F is an equilibrium state satisfying $H_i = 0$. The second derivative

$$\left.\frac{d^2h}{d\eta^2}\right|_F = H_{ss}(u_{s,\eta})^2 + H_{\alpha\alpha}(u_{\alpha,\eta})^2 \tag{1.5.3}$$

indicates that any ray in the sub-space spanned by the critical coordinates u_α ($\alpha = 1, 2, \ldots, n$) yields a zero second variation while an arbitrary ray outside this sub-space, involving at least one u_s, results in

$$\left.\frac{d^2h}{d\eta^2}\right|_F = H_{ss}(u_{s,\eta})^2 > 0, \tag{1.5.4}$$

and hence a positive second variation. It follows that higher variations of the energy function must now be examined with respect to an infinite

1.5.2 Stability of coincident critical points

number of paths initially lying in the n-dimensional u_α-space as opposed to the unique path tangential to the critical coordinate u_1 in the case of simple critical points (Section 1.1.4). It seems, therefore, that the higher variations of energy can best be examined with respect to u_α ($\alpha = 1, 2, \ldots, n$) directly by expressing u_s ($s = n+1, \ldots, N$) in terms of u_α [53], i.e., instead of the path (1.1.33), one considers the surface

$$u_i = u_i(u_\alpha) = u_{i,\alpha} u_\alpha + \tfrac{1}{2} u_{i,\alpha\beta} u_\alpha u_\beta + \cdots \tag{1.5.5}$$

in which $u_{\alpha,\alpha} \equiv du_\alpha/du_\alpha = 1$, $u_{\alpha,\beta} = u_{\alpha,\beta\gamma} = \cdots = 0$, and all $u_{s,\alpha} = 0$.

To examine the variations of energy with respect to (1.5.5), the change in energy is written in the form

$$h(u_\alpha) = H[u_i(u_\alpha), 0] - H(0, 0) \tag{1.5.6}$$

and differentiated with respect to u_α ($\alpha = 1, 2, \ldots, n$) to obtain, after evaluation

$$\left.\frac{\partial h}{\partial u_\alpha}\right|_F = H_i u_{i,\alpha} = 0$$

since $H_i = 0$.

The second differentiation with respect to u_β ($\beta = 1, 2, \ldots, n$) yields

$$\frac{\partial^2 h}{\partial u_\alpha \partial u_\beta} = H_{ij} u_{i,\alpha} u_{j,\beta} + H_i u_{i,\alpha\beta}$$

which, on evaluation, results in

$$\left.\frac{\partial^2 h}{\partial u_\alpha \partial u_\beta}\right|_F = H_{ss} u_{s,\alpha} u_{s,\beta} = 0 \tag{1.5.7}$$

since all $u_{s,\alpha} = 0$ along the surface (1.5.5).

The equation (1.5.7) confirms once more that if any $u_{s,\alpha} \neq 0$, the corresponding second variation

$$\delta^2 h = \frac{1}{2} \left.\frac{\partial^2 h}{\partial u_\alpha \partial u_\beta}\right|_F u_\alpha u_\beta$$

will be positive and the directions along which it vanishes are given by $u_{s,\alpha} = 0$ ($\alpha = 1, 2, \ldots, n$). (Note that the above quadratic form can be put in a form of sum of squares.)

Now evaluate the third variation under this condition. The third derivative is

$$\frac{\partial^3 h}{\partial u_\alpha \partial u_\beta \partial u_\gamma} = H_{ijk} u_{i,\alpha} u_{j,\beta} u_{k,\gamma} + H_{ij} u_{i,\alpha\gamma} u_{j,\beta}$$

$$+ H_{ij} u_{i,\alpha} u_{j,\beta\gamma} + H_{ij} u_{i,\alpha\beta} u_{j,\gamma} + H_i u_{i,\alpha\beta\gamma}$$

1.5 Coincident critical points

which yields

$$\left.\frac{\partial^3 h}{\partial u_\alpha \, \partial u_\beta \, \partial u_\gamma}\right|_F = H_{\alpha\beta\gamma} \tag{1.5.8}$$

It follows immediately that a necessary condition of stability is given by

$$H_{\alpha\beta\gamma} = 0; \quad \alpha, \beta, \gamma = 1, 2, \ldots, n \tag{1.5.9}$$

If this condition is satisfied the analysis is carried a step further to examine the fourth variation

$$\delta^4 h = \frac{1}{4!}[H_{\alpha\beta\gamma\delta} + 6H_{s\gamma\delta}u_{s,\alpha\beta} + 3H_{ss}u_{s,\alpha\beta}u_{s,\gamma\delta}]u_\alpha u_\beta u_\gamma u_\delta \tag{1.5.10}$$

which is obtained similarly by evaluating the fourth derivative of the energy at F.

It is seen once more that the second derivatives $u_{s,\alpha\beta}$ of the surface (1.5.5) are in the picture. In analogy with simple critical points, however, $u_{s,\alpha\beta}$ can readily be determined by minimizing the fourth variation of the energy with respect to these derivatives. Thus,

$$\frac{\partial(\delta^4 h)}{\partial(u_{r,\alpha\beta})} = \frac{1}{4!}(H_{r,\gamma\delta} + H_{rr}u_{r,\gamma\delta})u_\gamma u_\delta = 0 \tag{1.5.11}$$

in which summation on r is suspended, yielding

$$u_{r,\gamma\delta} = -\frac{H_{r,\gamma\delta}}{H_{rr}} \tag{1.5.12}$$

Using (1.5.10) and (1.5.12) one gets

$$\delta^4 h = \frac{1}{4!}\left(H_{\alpha\beta\gamma\delta} - 3\frac{H_{s\alpha\beta}H_{s\gamma\delta}}{H_{ss}}\right)u_\alpha u_\beta u_\gamma u_\delta \tag{1.5.13}$$

It is concluded that the fourth variation $\delta^4 h$ is positive for all $u_\alpha \neq 0$ if and only if the expression in the parentheses is positive definite. This is, then, a sufficient condition of stability.

If (1.5.13) admits negative values for some combinations of u_α, then, the state obviously is unstable. It might turn out that (1.5.13) is positive semi-definite, rendering an indecisive situation once more. The established procedure, however, can now readily be repeated and higher order variations be evaluated.

The surface derivatives $u_{s,\alpha}$, $u_{s,\alpha\beta}$, etc. can also be evaluated separately in advance of the stability analysis. Thus, following the method in Section 1.1.4, one substitutes the assumed solution $u_i(u_\alpha)$ into the

1.5.2 Stability of coincident critical points

necessary conditions for a minimum, $H_i = 0$, to get

$$H_i[u_j(u_\alpha), 0] \equiv 0 \tag{1.5.14}$$

Differentiating with respect to u_α ($\alpha = 1, 2, \ldots, n$) yields

$$H_{ij} u_{j,\alpha} = 0$$

which, on evaluation at F, yields

$$u_{s,\alpha} = 0 \quad \text{for} \quad s = n+1, \ldots, N, \tag{1.5.15}$$

the first derivatives of the surface (1.5.5) predicted earlier in the analysis. The second differentiation yields

$$H_{ijk} u_{j,\alpha} u_{k,\beta} + H_{ij} u_{j,\alpha\beta} = 0$$

which, in turn, results in

$$\begin{aligned} H_{\alpha\beta\gamma} &= 0 & \text{for} \quad i = \gamma \\ u_{s,\alpha\beta} &= -\frac{H_{s\alpha\beta}}{H_{ss}} & \text{for} \quad i = s \end{aligned} \tag{1.5.16}$$

the second surface derivatives (1.5.12), determined through other arguments earlier. The derivatives (1.5.16) also reveal that the second derivatives are only meaningful when all $H_{\alpha\beta\gamma} = 0$.

The analysis now simply consists of sequential perturbations of (1.5.6) and (1.5.14).

Obviously, a simple critical point can be considered as a special case of the more general coincident critical points in which $n = 1$. The analysis developed here, therefore, provides an all-embracing procedure for examining the stability of any critical point, irrespective of whether it is simple or coincident.

Returning now to the beginning of this section, it is worth pursuing the approach involving the arbitrary path (1.1.33) which was abandoned in favour of the surface (1.5.5); the reason for this is that, while the former approach is not as direct as the latter as far as the stability of F is concerned, it may, on the other hand, furnish information or some clue as to the equilibrium paths emanating from the coincident critical point which will form the subject of the following section.

In fact, the path derivatives $u_{s,1}$, $u_{s,11}$ in Section 1.1.4 are obtained by perturbing $H_i[u_j(u_1), 0] \equiv 0$ which represents the necessary conditions for a minimum while describing the equilibrium equations with a constant loading parameter. This has important practical implications since the derivatives of a post-buckling analysis (with $\Lambda = \Lambda^F$) can, then, be used in examining the stability of F or vice-versa.

1.5 Coincident critical points

Proceeding from (1.5.3), one observes that unless

$$u_{s,\eta} = 0 \qquad (1.5.17)$$

the second variation is positive, and the third variation must be examined for paths with $u_{s,\eta} = 0$.

Differentiating (1.5.2) for a third time with respect to η yields

$$\frac{d^3 h}{d\eta^3} = H_{ijk} u_{i,\eta} u_{j,\eta} u_{k,\eta} + 3 H_{ij} u_{i,\eta\eta} u_{j,\eta} + H_i u_{i,\eta\eta\eta}$$

and

$$\left.\frac{d^3 h}{d\eta^3}\right|_F = H_{\alpha\beta\gamma} u_{\alpha,\eta} u_{\beta,\eta} u_{\gamma,\eta} \qquad (1.5.18)$$

The paths that minimize (1.5.18) can be obtained (if there is any) from

$$\frac{\partial(\delta^3 h)}{\partial(u_{\alpha,\eta})} = H_{\alpha\beta\gamma} u_{\beta,\eta} u_{\gamma,\eta} = 0 \qquad (1.5.19)$$

which is in the form of n homogeneous equations. It will be seen in the next section that (1.5.19) can be obtained from the equilibrium equations (1.5.30) by setting $\Lambda = \Lambda^c$, resulting in $\Lambda_{,\eta} = 0$.

Suppose that (1.5.19) yields at least one real solution, (1.5.18) will then vanish for such a solution, but it will be generally non-zero for other paths unless all

$$H_{\alpha\beta\gamma} = 0 \quad \text{for} \quad \alpha, \beta, \gamma = 1, 2, \ldots, n, \qquad (1.5.20)$$

the necessary condition of stability derived earlier. If this condition is satisfied, the fourth derivative

$$\frac{d^4 h}{d\eta^4} = H_{ijkl} u_{i,\eta} u_{j,\eta} u_{k,\eta} u_{l,\eta} + 6 H_{ijk} u_{i,\eta\eta} u_{j,\eta} u_{k,\eta}$$
$$+ 3 H_{ij} u_{i,\eta\eta} u_{j,\eta\eta} + 4 H_{ij} u_{i,\eta\eta\eta} u_{j,\eta} + H_i u_{i,\eta\eta\eta\eta}$$

yields

$$\left.\frac{d^4 h}{d\eta^4}\right|_F = H_{\alpha\beta\gamma\delta} u_{\alpha,\eta} u_{\beta,\eta} u_{\gamma,\eta} u_{\delta,\eta} + 6 H_{s\alpha\beta} u_{s,\eta\eta} u_{\alpha,\eta} u_{\beta,\eta} + 3 H_{ss} u_{s,\eta\eta}^2 \qquad (1.5.21)$$

The particular $u_{s,\eta\eta}$ along which the change in energy is minimum can be obtained from

$$\frac{\partial(\delta^4 h)}{\partial(u_{s,\eta\eta})} = H_{s\alpha\beta} u_{\alpha,\eta} u_{\beta,\eta} + H_{ss} u_{s,\eta\eta} = 0$$

as

$$u_{s,\eta\eta} = -\frac{H_{s\alpha\beta} u_{\alpha,\eta} u_{\beta,\eta}}{H_{ss}}$$

1.5.2 Stability of coincident critical points

which, upon substitution back into (1.5.21), yields

$$\left.\frac{d^4 h}{d\eta^4}\right|_F = \frac{1}{4!}\left(H_{\alpha\beta\gamma\delta} - 3\frac{H_{s\alpha\beta}H_{s\gamma\delta}}{H_{ss}}\right)u_{\alpha,\eta}u_{\beta,\eta}u_{\gamma,\eta}u_{\delta,\eta} \qquad (1.5.22)$$

Obviously, if the expression in parentheses is positive definite, then, $d^4 h/d\eta^4|_F > 0$ for all rays described by η and F is stable.

Determination of the path derivatives in advance as before, also provides some insight into the nature of the path. Thus, differentiating

$$H_i[u_j(\eta), 0] \equiv 0 \qquad (1.5.23)$$

successively yields

$$H_{ij}u_{j,\eta} = 0$$

and upon evaluation

$$u_{s,\eta} = 0.$$

Similarly

$$H_{ijk}u_{j,\eta}u_{k,\eta} + H_{ij}u_{j,\eta\eta} = 0$$

which results in

$$H_{\alpha\beta\gamma}u_{\beta,\eta}u_{\gamma,\eta} = 0 \quad \text{for} \quad i = \alpha \qquad (1.5.24)$$

and

$$u_{s,\eta\eta} = -\frac{H_{s\alpha\beta}u_{\alpha,\eta}u_{\beta,\eta}}{H_{ss}} \quad \text{for} \quad i = s \qquad (1.5.25)$$

Thus, previously derived path derivatives are recovered. (1.5.24) supplies n homogeneous equations for the derivatives $u_{\alpha,\eta}$ defining the positions of the paths in the u_α-space, along which the change in energy is minimum unless all $H_{\alpha\beta\gamma} = 0$ resulting in an infinite number of such paths. It must be noted that $u_{s,\eta\eta}$ given by (1.5.25) is only compatible with such paths, and even more significantly, if any $H_{\alpha\beta\gamma} \neq 0$ the solution rays of (1.5.24) do not play a central role in the stability of F since the fact that some $H_{\alpha\beta\gamma} \neq 0$ renders F unstable. In other words, following the directions obtained from (1.5.24) when $H_{\alpha\beta\gamma} \neq 0$ and determining the fourth derivative of h does not lead to a decision on stability in contrast with simple critical points, and the motivation of introducing the surface (1.5.5) to replace the path described by η is now clearer. The analysis involving the path, however, will provide cross-reference between the present and following sections.

1.5 Coincident critical points

1.5.3 Equilibrium paths emanating from a coincident critical point: compound branching

Section 1.1.5 supplies a general systematic perturbation method for obtaining ordered estimates of primary critical loads at which an initially stable and, in general, nonlinear fundamental path loses its stability. Neither in Section 1.1.5, nor in the corresponding analysis of symmetric systems in Section 1.3.4, is there an assumption that would limit these techniques to simple critical points, and the critical loads associated with coincident points, therefore, can similarly be obtained.

The post-critical behaviour in the vicinity of a coincident critical point, however, presents complexities as remarked earlier, and will now be discussed.

The function $S(u_i, \Lambda)$ of Section 1.1.6, with all its essential properties, can be used here once more, and it is implicit, therefore, that attention will be focussed on branching phenomena.

Assume now that C is a coincident critical point on the fundamental path at which

$$S_{\alpha\alpha}(0, \Lambda^c) = 0 \qquad \alpha = 1, 2, \ldots, n; \quad n < N$$

and (1.5.26)

$$S_{ss}(0, \Lambda^c) \neq 0 \qquad s = n+1, \ldots, N;$$

implying that u_α and u_s denote critical and non-critical coordinates respectively. It must be remarked immediately that u_α does not represent a unique subset of principal coordinates at C. It is well-known that coincident eigenvalues can have an infinite number of principal directions, and, in fact, any linear combination of u_α indicates a principal direction. The widely used phrase 'coupled modes of buckling' referring to post-buckling paths appears, therefore, to be meaningless since there are no well-defined modes of buckling at a coincident critical point.

Assume now that any post-buckling path emanating from C is expressed in the parametric form

$$u_i = u_i(\eta), \qquad \Lambda = \Lambda(\eta) \tag{1.5.27}$$

where η is an appropriate perturbation parameter. As remarked earlier, apart from the fundamental path $u_i = 0$, there may be several post-buckling paths, and it is important to realize at this stage that each of these paths may require a different parameter for adequate representation. More specifically, it is now possible to have a path tangential to the u_2 axis (say), thus ruling out u_1 as a perturbation parameter and vice-versa. Keeping η, therefore, unspecified for now, substitute the assumed

1.5.3 Equilibrium paths: compound branching

solutions (1.5.27) into the equilibrium equations $S_i = 0$ to get the identities

$$S_i[u_j(\eta), \Lambda(\eta)] \equiv 0 \tag{1.5.28}$$

which upon differentiation with respect to η yields

$$S_{ij}u_{j,\eta} + S'_i\Lambda_\eta = 0$$

and evaluation at C results in

$$u_{s,\eta} = 0 \quad \text{for} \quad i = s = n+1, \ldots, N \tag{1.5.29}$$

It is noted that (1.5.29) is identical with (1.5.17) which is associated with the path along which the change in energy is minimum.

A second perturbation leads to

$$(S_{ijk}u_{k,\eta} + S'_{ij}\Lambda_\eta)u_{j,\eta} + S_{ij}u_{j,\eta\eta} + (S'_{ij}u_{j,\eta} + S''_i\Lambda_\eta)\Lambda_\eta + S'_i\Lambda_{\eta\eta} = 0$$

which results in

$$S_{\alpha\beta\gamma}u_{\beta,\eta}u_{\gamma,\eta} + 2S'_{\alpha\alpha}\Lambda_\eta u_{\alpha,\eta} = 0 \quad \text{for} \quad i = \alpha \tag{1.5.30}$$

and

$$u_{s,\eta\eta} = -\frac{S_{s\alpha\beta}u_{\alpha,\eta}u_{\beta,\eta}}{S_{ss}} \quad \text{for} \quad i = s \tag{1.5.31}$$

It is again observed that (1.5.31) corresponds to (1.5.25).

Supposing now that one of the basic variables, Λ say, can describe all possible paths, and setting $\eta = \Lambda$, it is observed that (1.5.30) represents a set of n second order equations in n unknowns, $u_{\alpha,\eta} \equiv u'_\alpha$ which yields $2 \cdot 2 \cdot 2 \cdots 2 = 2^n$ sets of solutions (solution rays) in the real or complex field provided the set (1.5.30) does not have an identically zero Jacobian,

$$J = \det \begin{vmatrix} \dfrac{\partial f_1}{du'_1} & \dfrac{\partial f_1}{\partial u'_2} & \dfrac{\partial f_1}{\partial u'_n} \\ \cdots & \cdots & \cdots \\ \cdots & \cdots & \cdots \\ \dfrac{\partial f_n}{\partial u'_1} & \dfrac{\partial f_n}{\partial u'_2} & \dfrac{\partial f_n}{\partial u'_n} \end{vmatrix} \neq 0 \tag{1.5.32}$$

where f_α ($\alpha = 1, 2, \ldots, n$) denotes the polynomial of the corresponding equation in the set (1.5.30). The condition (1.5.32) ensures that the equations in the set (1.5.30) are *independent*; otherwise one solution follows from another one and the system has infinitely many solutions.

If the fundamental path is excluded, then, *the number of post-buckling solutions is $2^n - 1$.*

The choice of Λ as the parameter η here does not represent a loss of

1.5 Coincident critical points

generality. In fact, if one chooses an appropriate combination of basic variables u_α and Λ as the parameter η, then, an additional relationship between these variables and η comes into the picture, thus resulting in $n+1$ equations in $n+1$ unknowns. Consider, for example, the ray

$$u_\alpha = l_\alpha \eta$$
$$\Lambda = l\eta \qquad (1.5.33)$$

in the $u_\alpha - \Lambda$ space, where l_α and l are the direction cosines, and η measures the distance from the origin. One can associate each solution ray $(u_{\alpha,\eta}, \Lambda_\eta)$ of (1.5.30) with a ray of the form (1.5.33) which is actually tangential to the complete solution path. It follows from (1.5.33) and the definition of direction cosines that

$$\sum_{\alpha=1}^{n} u_{\alpha,\eta}^2 + \Lambda_\eta^2 = 1 \qquad (1.5.34)$$

Thus, the problem of finding the solution rays takes the form of finding the sets of direction cosines $(u_{\alpha,\eta}, \Lambda_\eta)$. One now has $n+1$ equations, (1.5.30) and (1.5.34), and $n+1$ unknowns. Each equation has a polynomial of *degree* 2 and the number of solution rays, therefore, appears to be 2^{n+1} provided the Jacobian is not identically zero. However, it is observed that if $(u_{\alpha,\eta}, \Lambda_\eta)$ is a solution ray, then $(-u_{\alpha,\eta}, -\Lambda_\eta)$ is also a solution ray and the actual number of the rays is, therefore, given by $\frac{1}{2}2^{n+1} = 2^n$ as predicted earlier.

In [76] a method leading to the selection of optimal perturbation parameters is discussed extensively.

In the stability theory one is naturally concerned with real solutions of (1.5.30). Since complex roots occur in pairs, and due to the fact that a real path (fundamental) is already known, there must be at least *one* post-buckling path emanating from C [70]. Consider, for example, the particular case in which $n=1$ and C is simple; setting $\eta = \Lambda$ in (1.5.30), then, yields

$$(S_{111} u_1' + 2S_{11}')u_1' = 0$$

which results in the fundamental path

$$u_1' = 0$$

and the post-buckling path

$$u_1' = -\frac{2S_{11}'}{S_{111}}$$

in which the energy derivatives are assumed to be finite. Note that the

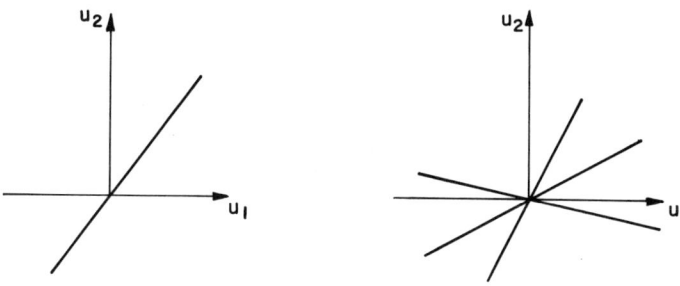

Figure 1.25

latter result can also be obtained by setting $\eta = u_1$ in (1.5.30), as

$$\Lambda_1 = -\frac{S_{111}}{2S'_{11}},$$

but the fundamental path $u_1 = 0$ is, then, missed since u_1 is sliding along the path.

Similarly, for $n = 2$ there is at least *one* and possibly *three* postbuckling paths (Figure 1.25), for $n = 3$ a maximum of *seven* and in general a maximum of $2^n - 1$ paths.

Finally, it is observed that the curvatures $u_{s,\eta\eta}$ of these paths are readily evaluated by substituting for $u_{\alpha,\eta}$ into (1.5.31).

1.5.4 Symmetric systems

Suppose now that the system exhibits symmetry in some or all critical coordinates such that

$$S(u_s, u_\alpha, z_\beta, \Lambda) = S(u_s, u_\alpha, -z_\beta, \Lambda) \tag{1.5.35}$$

as in Section 1.3.2, where z_β ($\beta = 1, 2, \ldots, m$; $m \leq n$) distinguishes the symmetric coordinates from the remaining critical coordinates u_α ($\alpha = m+1, \ldots, n$), and all z_β *change sign as a set*. In this section, however, symmetry of S in z_β *individually* will also be considered. The latter symmetry is, of course, stronger than the former, and eliminates terms such as $S_{s12} \equiv \dfrac{\partial^3 S}{\partial u_s\, \partial z_1\, \partial z_2}$ which can appear when z_1 and z_2 change sign together. In the sequel, individual symmetry will be indicated *whenever* it is imposed, otherwise the *set-symmetry* as in Section 1.3.2 will be implied.

It will be instructive to consider first some special cases. For instance, let $m = 1$, then, (1.5.30) yields for $\alpha = 1$:

$$S_{11\gamma}z_{1,\eta}u_{\gamma,\eta} + S'_{11}\Lambda_\eta z_{1,\eta} = 0 \qquad (\gamma = 2, 3, \ldots, n) \tag{1.5.36}$$

1.5 Coincident critical points

which, in turn, results in

$$z_{1,\eta} = 0 \tag{1.5.37}$$

or

$$S_{11\gamma}u_{\gamma,\eta} + S'_{11}\Lambda_\eta = 0 \tag{1.5.38}$$

Similarly, for $\alpha = 2, 3, \ldots, n$, (1.5.30) yields

$$S_{\alpha 11}z_{1,\eta}^2 + S_{\alpha\beta\gamma}u_{\beta,\eta}u_{\gamma,\eta} + 2S'_{\alpha\alpha}\Lambda_\eta u_{\alpha,\eta} = 0 \tag{1.5.39}$$

where $\alpha, \beta, \gamma = 2, 3, \ldots, n$.

It is observed that if z_1 were chosen as the perturbation parameter η in advance, the path associated with (1.5.37) would have been missed. Obviously, this path has no projection on the z_1 axis, and it is further deduced that if there are more symmetric coordinates, for each z_β ($\beta = 1, 2, \ldots, m$) there can be such a path with $z_{\beta,\eta} = 0$ provided *individual symmetry* applies and $m < n$. Furthermore, if $(z_{1,\eta}, u_{\alpha,\eta}, \Lambda_\eta)$ satisfies (1.5.39) so does $(-z_{1,\eta}, u_{\alpha,\eta}, \Lambda_\eta)$ indicating a symmetric behaviour with respect to z_1. Assuming further that $n = 2$ while $m = 1$ and setting $\eta = u_2$ one gets more explicit results,

$$z_{1,2} = 0, \qquad \Lambda_2 = -\frac{S_{222}}{2S'_{22}}$$

$$\Lambda_2 = -\frac{S_{211}}{S'_{11}}, \qquad (z_{1,2})^2 = -\frac{1}{S_{211}}\left(S_{222} - 2S_{211}\frac{S'_{22}}{S'_{11}}\right) \tag{1.5.40}$$

(1.5.40) represents *one* or *three* solution rays depending on whether $(z_{1,2})^2$ is negative or positive respectively. Figures 1.26 and 1.27 show the rays in $z_1 - u_2$ and $\Lambda - u_2$ planes. Evidently, despite symmetry in the z_1 coordinate, there is no post-buckling path with a zero slope and the behaviour is asymmetric in the traditional sense as opposed to that

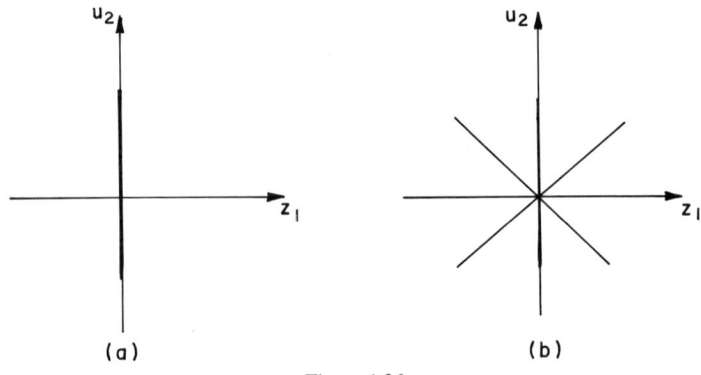

Figure 1.26

1.5.4 Symmetric systems

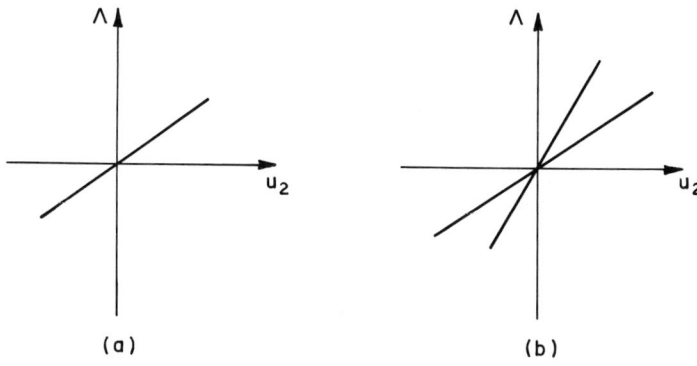

Figure 1.27

associated with simple critical points, i.e., $S_{111} = 0$ here does not result in a symmetric post-buckling path.

It was shown in the preceding section that unless all $S_{\alpha\beta\gamma} = 0$, the critical point is unstable, and one observes now that the post-buckling paths generally have finite slopes. The next question of interest is therefore concerned with the case in which all $S_{\alpha\beta\gamma} = 0$. Thus, setting $n = m$ results in all $S_{\alpha\beta\gamma} = 0$ on the basis of the symmetry condition (1.5.35), and the equation (1.5.30) then, yields

$$\Lambda_\eta = 0 \tag{1.5.41}$$

for post-buckling paths which seems to indicate that any $u_{\alpha,\eta}$ satisfies (1.5.30). However, one must not rush to conclusions before examining higher order equations.

Differentiating (1.5.28) with respect to η for a third time and evaluating under the condition that all $S_{\alpha\beta\gamma} = 0$ results in

$$S_{\alpha\beta\gamma\delta}z_{\beta,\eta}z_{\gamma,\eta}z_{\delta,\eta} + 3S_{s\alpha\beta}z_{\beta,\eta}u_{s,\eta\eta}$$
$$+ 3S'_{\alpha\alpha}\Lambda_{\eta\eta}z_{\alpha,\eta} = 0, \quad \text{for} \quad i = \alpha \tag{1.5.42}$$

Substituting for $u_{s,\eta\eta}$ from (1.5.31) yields

$$\left(S_{\alpha\beta\gamma\delta} - 3\frac{S_{s\alpha\beta}S_{s\gamma\delta}}{S_{ss}}\right)z_{\beta,\eta}z_{\gamma,\beta}z_{\delta,\eta} + 3S'_{\alpha\alpha}\Lambda_{\eta\eta}z_{\alpha,\eta} = 0 \tag{1.5.43}$$

Obviously, if $(z_{\alpha,\eta}, \Lambda_{\eta\eta})$ is a solution so is $(-z_{\alpha,\eta}, \Lambda_{\eta\eta})$, indicating symmetry around Λ. It may, therefore, be expected that (1.5.43) in conjunction with (1.5.34) which is now in the form

$$\sum_{\alpha=1}^{n} z_{\alpha,\eta}^2 = 1 \tag{1.5.44}$$

yields $\frac{1}{2} \cdot 3^n = 3^n$ solutions [76] in real or complex domain.

1.5 Coincident critical points

Assuming, however, that *individual symmetry* holds, it was first shown by Chilver and Johns [73] that the maximum number of solutions is given by $\frac{1}{2}(3^m - 1)$. Individual symmetry reduces (1.5.43) to the form

$$\bar{S}_{\alpha\alpha\alpha\alpha}z_{\alpha,\eta}^3 + 3\bar{S}_{\alpha\alpha\beta\beta}z_{\alpha,\eta}z_{\beta,\eta}z_{\beta,\eta} + 3S'_{\alpha\alpha}\Lambda_{\eta\eta}z_{\alpha,\eta} = 0 \qquad (1.5.45)$$

where

$$\bar{S}_{\alpha\alpha\beta\beta} = S_{\alpha\alpha\beta\beta} - 3\frac{S_{s\alpha\alpha}S_{s\beta\beta}}{S_{ss}} \quad \text{and} \quad \alpha \neq \beta.$$

It is worth noting that summation convention works efficiently here as well as before. Thus one observes that while there is no summation over α (since double α on S, for instance, cancels it), summation is implied on β since two β's follow the double β on $\bar{S}$ activating the convention. This of course has been implied all along as in the case of $\frac{1}{S_{ss}}S_{s\alpha\alpha}S_{s\beta\beta}$ which indicates summation over s.

Obviously (1.5.45) has at least n solutions in the form

$$\Lambda_{\eta\eta} = -\frac{\bar{S}_{\alpha\alpha\alpha\alpha}z_{\alpha,\eta}^2}{3S'_{\alpha\alpha}}, \quad \text{all} \quad z_{\beta,\eta} = 0 \ (\alpha \neq \beta) \qquad (1.5.46)$$

where α takes a particular value from 1 to n, i.e., there is one post-buckling path along each axis z_α. Armed with this knowledge, in simple cases (e.g., when $m = 2$) one can determine other solution rays conveniently by assigning the role of η to one of the critical coordinates without missing a path. In more complex situations the role of η can be assigned to each one of the critical coordinates in turn since the resulting multiplication of some solutions can readily be identified.

Having determined the solution rays $z_{\alpha,\eta}$ the corresponding curvatures $u_{s,\eta\eta}$ and $\Lambda_{\eta\eta}$ can readily be obtained from (1.5.31) and (1.5.43) respectively. Figure 1.28 shows some of the paths for $m = 2$.

Evidently, the critical point C is a symmetric point of bifurcation in

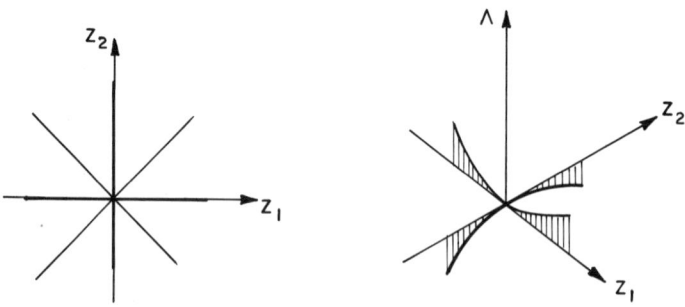

Figure 1.28

1.5.5 Imperfect systems

the sense that all post-buckling paths have a zero slope; the curvatures $\Lambda_{\eta\eta}$, however, do not necessarily have the same sign and whether C is a stable-symmetric or unstable-symmetric point might seem to be uncertain. One recalls now the stability discussion concerning the point C in the preceding section; it was shown there that C is stable if the expression in parentheses in (1.5.43) or equivalently in (1.5.22) is positive definite. If this condition is not satisfied, the critical point C is unstable even if some post-buckling paths may have a positive curvature.

1.5.5 Imperfect systems

The behaviour of a slightly imperfect structure is bound to be complex, if the system, in its perfect form, exhibits several post-buckling paths passing through a coincident critical point. A variety of behaviour can be expected depending on various factors such as the nature of imperfections, slopes of post-buckling paths, stability of the critical point, etc. It seems, therefore, that it may be more fruitful to examine individual physical systems rather than attempting to develop a general theory. Nevertheless, it is possible to acquire a valuable insight into many important aspects of imperfection-sensitivity associated with coincident critical points by considering certain limited cases in the context of the general theory.

Consider once more the system described by the potential energy function $S(u_i, \Lambda)$ as in Sections 1.1.6 and 1.5.3 and assume as in the latter section that stability is lost at a coincident critical point C given by (1.5.26). In analogy with simple critical points, a family of imperfect systems can be generated by introducing a small imperfection parameter, ϵ, into the system. Equilibrium equations of resulting imperfect systems, $S_i(u_j, \Lambda, \epsilon) = 0$, can then be solved simultaneously to yield solutions in the parametric form

$$u_i = u_i(\eta_1, \eta_2), \qquad \Lambda = \Lambda(\eta_1, \eta_2), \qquad \epsilon = \epsilon(\eta_1, \eta_2) \qquad (1.5.47)$$

where η_1 and η_2 are two *independent* perturbation parameters. To investigate the equilibrium surface in the vicinity of C, the assumed solutions (1.5.47) are substituted back into the equilibrium equations $S_i = 0$ to get the identities

$$S_i[u_j(\eta_1, \eta_2), \Lambda(\eta_1, \eta_2), \epsilon(\eta_1, \eta_2)] \equiv 0 \qquad (1.5.48)$$

Differentiating these identities first with respect to η_1 and then with respect to η_2 results in

$$S_{ij}u_{j,1} + S'_i\Lambda_1 + \dot{S}_i\epsilon_1 = 0 \qquad (1.5.49)$$

1.5 Coincident critical points

and
$$S_{ij}u_{j,2} + S'_i\Lambda_2 + \dot{S}_i\epsilon_2 = 0 \tag{1.5.50}$$

respectively, where subscripts 1 and 2 denote partial differentiation with respect to η_1 and η_2.

It will be assumed here that ϵ is associated with one of the critical coordinates, u_1 say, only such that

$$\dot{S}_1 \equiv \left.\frac{\partial S_1}{\partial \epsilon}\right|_C \neq 0 \quad \text{while} \quad \dot{S}_\delta = 0 \ (\delta = 2, 3, \ldots, n) \tag{1.5.51}$$

and in general $\dot{S}_s \neq 0$ $(s = n+1, \ldots, N)$.

This is not a severe restriction as it may seem to be, considering the arbitrariness in orientation of u_α axes in space with respect to u_s axes which stems from the coincidence of eigenvalues. In fact, it is possible to introduce an appropriate transformation of u_α based on the multiplicity of the eigenvalue $S_{\alpha\alpha}$ such that all $\dot{S}_\delta$ ($\delta \neq 1$) vanish while $\dot{S}_1 \neq 0$ [79].

Evaluating now (1.5.49) and (1.5.50) at C results in

$$\epsilon_1 = 0 \quad \text{for} \quad i = 1, \quad u_{s,1} = 0 \quad \text{for} \quad i = s \tag{1.5.52}$$

and
$$\epsilon_2 = 0 \quad \text{for} \quad i = 1, \quad u_{s,2} = 0 \quad \text{for} \quad i = s \tag{1.5.53}$$

A second perturbation yields

$$(S_{ijk}u_{k,1} + S'_{ij}\Lambda_1 + \dot{S}_{ij}\epsilon_1)u_{j,1} + S_{ij}u_{j,11}$$
$$+ (S'_{ij}u_{j,1} + S''_i\Lambda_1 + \dot{S}'_i\epsilon_1)\Lambda_1 + S'_i\Lambda_{11}$$
$$+ (\dot{S}_{ij}u_{j,1} + \dot{S}'_i\Lambda_1 + \ddot{S}_i\epsilon_1)\epsilon_1 + \dot{S}_i\epsilon_{11} = 0, \tag{1.5.54}$$

$$(S_{ijk}u_{k,2} + S'_{ij}\Lambda_2 + \dot{S}_{ij}\epsilon_2)u_{j,1} + S_{ij}u_{j,12}$$
$$+ (S'_{ij}u_{j,2} + S''_i\Lambda_2 + \dot{S}'_i\epsilon_2)\Lambda_1 + S'_i\Lambda_{12}$$
$$+ (\dot{S}_{ij}u_{j,2} + \dot{S}'_i\Lambda_2 + \ddot{S}_i\epsilon_2)\epsilon_1 + \dot{S}_i\epsilon_{12} = 0, \tag{1.5.55}$$

and
$$(S_{ijk}u_{k,2} + S'_{ij}\Lambda_2 + \dot{S}_{ij}\epsilon_2)u_{j,2} + S_{ij}u_{j,22}$$
$$+ (S'_{ij}u_{j,2} + S''_i\Lambda_2 + \dot{S}'_i\epsilon_2)\Lambda_2 + S'_i\Lambda_{22}$$
$$+ (\dot{S}_{ij}u_{j,2} + \dot{S}'_i\Lambda_2 + \ddot{S}_i\epsilon_2)\epsilon_2 + \dot{S}_i\epsilon_{22} = 0 \tag{1.5.56}$$

Evaluations at C result in a *split between the critical and noncritical subspaces*;

$$\left.\begin{array}{l}S_{\alpha\beta\gamma}u_{\beta,1}u_{\gamma,1} + 2S'_{\alpha\alpha}\Lambda_1 u_{\alpha,1} + \dot{S}_\alpha\epsilon_{11} = 0 \\ S_{s\beta\gamma}u_{\beta,1}u_{\gamma,1} + S_{ss}u_{s,11} + \dot{S}_s\epsilon_{11} = 0\end{array}\right\} \tag{1.5.57}$$

$$\left.\begin{array}{l}S_{\alpha\beta\gamma}u_{\beta,2}u_{\gamma,1} + S'_{\alpha\alpha}\Lambda_2 u_{\alpha,1} + S'_{\alpha\alpha}\Lambda_1 u_{\alpha,2} + \dot{S}_\alpha\epsilon_{12} = 0 \\ S_{s\beta\gamma}u_{\beta,2}u_{\gamma,1} + S_{ss}u_{s,12} + \dot{S}_s\epsilon_{12} = 0\end{array}\right\} \tag{1.5.58}$$

1.5.5 Imperfect systems

and

$$S_{\alpha\beta\gamma}u_{\beta,2}u_{\gamma,2} + 2S'_{\alpha\alpha}\Lambda_2 u_{\alpha,2} + \dot{S}_\alpha \epsilon_{22} = 0$$
$$S_{s\beta\gamma}u_{\beta,2}u_{\gamma,2} + S_{ss}u_{s,22} + \dot{S}_s \epsilon_{22} = 0 \tag{1.5.59}$$

It is understood that in these equations $\dot{S}_\alpha = 0$ for $\alpha \neq 1$.

In this generality, one cannot infer much about the behaviour of the imperfect system from the equations (1.5.57), (1.5.58) and (1.5.59). In order to gain some insight, let $n = 2$ and $S_{211} = 0$. This is equivalent to assuming that one of the post-buckling paths of the perfect system is in the u_1 direction as can readily be verified by solving (1.5.30) for $n = 2$ (Figure 1.29). Recalling that ϵ is also associated with u_1, the specialized system under consideration emerges as one which has imperfections in one of the *actual* buckling modes.

The critical coordinate u_1 and the loading parameter Λ can, then, be given the roles of η_1 and η_2 respectively, and the equations (1.5.57), (1.5.58) and (1.5.59) take the more explicit forms

$$\left.\begin{array}{l} S_{111} + S_{122}(u_{2,1})^2 + \dot{S}_1 \epsilon_{11} = 0 \\ 2S_{221}u_{2,1} + S_{222}(u_{2,1})^2 = 0 \\ S_{s11} + 2S_{s12}u_{2,1} + S_{s22}(u_{2,1})^2 + S_{ss}u_{s,11} + \dot{S}_s \epsilon_{11} = 0 \end{array}\right\} \tag{1.5.60}$$

$$\left.\begin{array}{l} S_{122}u_{2,1}u'_2 + S'_{11} + \dot{S}_1 \epsilon'_1 = 0 \\ S_{221}u'_2 + S_{222}u_{2,1}u'_2 + S'_{22}u_{2,1} = 0 \\ S_{s21}u'_2 + S_{s22}u_{2,1}u'_2 + S_{ss}u'_{s,1} + \dot{S}_s \epsilon'_1 = 0 \end{array}\right\} \tag{1.5.61}$$

and

$$\left.\begin{array}{l} S_{122}(u'_2)^2 + \dot{S}_1 \epsilon'' = 0 \\ S_{222}(u'_2)^2 + 2S'_{22}u'_2 = 0 \\ S_{s22}(u'_2)^2 + S_{ss}u''_s + \dot{S}_s \epsilon'' = 0 \end{array}\right. \tag{1.5.62}$$

respectively. Here a prime on the variables indicates that η_2 has been replaced by Λ.

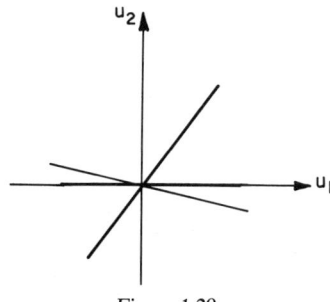

Figure 1.29

1.5 Coincident critical points

It is first observed that the second derivatives $u_{s,11}$, $u'_{s,1}$, u''_s associated with noncritical coordinates, can be obtained from the last set of equations in each group above, once the derivatives $u_{2,1}$, u'_2, ϵ_{11}, ϵ'_1 and ϵ'' are determined. The remaining 6 equations which are related to $u_\alpha - \epsilon$ subspace yield these latter derivatives in pairs, representing the two distinct solutions

$$u'_2 = 0, \quad u_{2,1} = 0, \quad \epsilon_{11} = -\frac{S_{111}}{\dot{S}_1}$$

$$\epsilon'_1 = -\frac{S'_{11}}{\dot{S}_1}, \quad \epsilon'' = 0 \tag{1.5.63}$$

and

$$u'_2 = -\frac{2S'_{22}}{S_{222}}, \quad u_{2,1} = -\frac{2S_{221}}{S_{222}},$$

$$\epsilon_{11} = -\frac{1}{\dot{S}_1}\left[S_{111} + 4\frac{(S_{221})^3}{(S_{222})^2}\right]$$

$$\epsilon'_1 = -\frac{1}{\dot{S}_1}\left[S'_{11} + 4\frac{S'_{22}(S_{221})^2}{(S_{222})^2}\right] \tag{1.5.64}$$

$$\epsilon'' = -\frac{4}{\dot{S}_1}\frac{S_{122}(S'_{22})^2}{(S_{222})^2}$$

The corresponding noncritical derivatives $u_{s,11}$, $u'_{s,1}$, u''_s follow immediately from the last sets of (1.5.60), (1.5.61) and (1.5.62) respectively; these derivatives, however, do not play an active role in characterizing the behaviour and it suffices here to note merely the derivatives of the first solution (1.5.63)

$$u_{s,11} = -\frac{1}{S_{ss}}\left(S_{s11} - S_{111}\frac{\dot{S}_s}{\dot{S}_1}\right)$$

$$u'_{s,1} = \frac{S'_{11}}{S_{ss}}\frac{\dot{S}_s}{\dot{S}_1} \quad u''_s = 0. \tag{1.5.65}$$

In the $u_\alpha - \lambda - \epsilon$ subspace, Taylor's theorem yields the two equilibrium surfaces

$$u_2 = u_2(u_1, \Lambda) = 0$$

$$\epsilon = \epsilon(u_1, \Lambda) = -\frac{1}{2\dot{S}_1}(S_{111}u_1^2 + 2S'_{11}u_1\lambda) \tag{1.5.66}$$

and

$$u_2 = -\frac{2}{S_{222}}(S_{221}u_1 + S'_{22}\lambda)$$

$$\epsilon = -\frac{1}{2\dot{S}_1}\left\{\left[S_{111} + 4\frac{(S_{221})^3}{(S_{222})^2}\right]u_1^2 + 2\left[S'_{11} + 4\frac{S'_{22}(S_{221})^2}{(S_{222})^2}\right]u_1\lambda + 4\frac{S_{221}(S'_{22})^2}{(S_{222})^2}\lambda^2\right\}$$

where $\lambda = \Lambda - \Lambda^c$. \hfill (1.5.67)

1.5.5 Imperfect systems

Interestingly, the first solution (1.5.66) is identical to the equilibrium surface (1.2.9) in the vicinity of a simple critical point, and the derivatives (1.5.65) confirm this fact. The imperfect system, then, deforms initially in $u_1 - \Lambda - \epsilon$ subspace following a certain path on the equilibrium surface (1.5.66), depending on ϵ; unlike the case of simple critical points, however, this path may lose its stability before reaching the limit point upon intersecting another path associated with the second solution (1.5.67).

Consider the intersection of the two equilibrium surfaces (1.5.66) and (1.5.67); in the $u_1 - \lambda$ subspace, the intersection takes the form of the straight line

$$S_{221}u_1 + S'_{22}\lambda = 0 \qquad (1.5.68)$$

which is obtained by setting $u_2 = 0$ in (1.5.67). The equilibrium path corresponding to a specified ϵ on the surface (1.5.66) may be intersected by the line (1.5.68) before the limit point, inducing bifurcation buckling with deflections developing in the u_2 direction. Figure 1.30 shows the line (1.5.68) lying between the asymptotes of the solution (1.5.66) for $\epsilon =$ constant, and intersecting the latter before the limit point. If the intersection takes place after the limit point or the line (1.5.68) lies outside the asymptotes, then, the system loses its stability at the limit point just like in the case of simple critical points. When the line (1.5.68) lies outside the asymptotes, however, it intersects the initially stable path associated with $-\epsilon$, resulting in bifurcation buckling again (Figure 1.31).

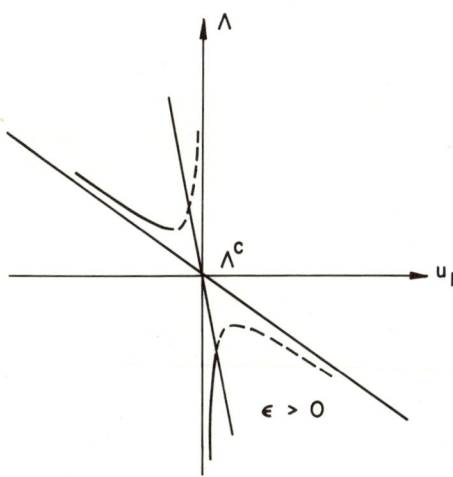

Figure 1.30

1.5 Coincident critical points

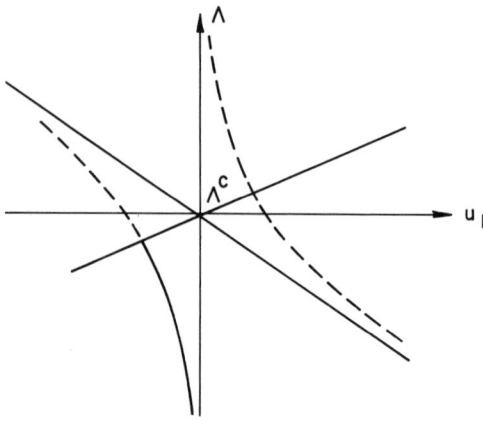

Figure 1.31

One observes that the behaviour of a slightly imperfect system in the vicinity of a coincident critical point can be quite different from that associated with simple points. Firstly, in contrast to simple critical points (asymmetric), there may not be a totally stable path in the vicinity of a coincident critical point, i.e., the equilibrium paths corresponding to $\epsilon > 0$ and $\epsilon < 0$ may both lose stability at a certain level of loading. The loss of stability may occur at a limit or bifurcation point in the former case (Figure 1.30) while it has to occur at a point of bifurcation in the latter. Secondly, the loss of stability may take place at much lower bifurcation loads, thus indicating higher imperfection sensitivity in the vicinity of coincident points.

It is evident from the foregoing discussion that critical load-imperfection relationships can readily be obtained from the above equilibrium analysis without necessarily undertaking a stability analysis. To this end, one first observes that the parabolic relationship (1.2.27),

$$\overset{*}{\Lambda} = \Lambda^c \pm \frac{(2S_{111}\dot{S}_1\epsilon)^{\frac{1}{2}}}{S'_{11}} \tag{1.5.69}$$

is still a possibility, which can also readily be derived here by locating the extrema on the surface (1.5.66). On the other hand, the bifurcation load-imperfection relationship can be obtained from the intersections discussed above, and in fact substituting for u_1 into (1.5.66) from (1.5.68) results in

$$\overset{*}{\lambda} = \pm \frac{S_{221}}{S'_{22}} \left[\frac{2\dot{S}_1\epsilon}{S_{111}\left(2\dfrac{S'_{11}S_{221}}{S_{111}S'_{22}} - 1\right)} \right]^{\frac{1}{2}} \tag{1.5.70}$$

1.5.5 Imperfect systems

where $\overset{*}{\lambda} = \overset{*}{\Lambda} - \Lambda^c$. Since the result (1.5.70) represents the intersection of the surfaces (1.5.66) and (1.5.67), it must also be obtainable through elimination of u_1 between (1.5.68) and the second equation in (1.5.67), and if this operation is carried out, (1.5.70) is indeed recovered.

It is noted that the relationship (1.5.70) is again *parabolic*; the points of the parabola, however, are now associated with bifurcation points. Supposing that $S'_{11} < 0$, $S_{111} < 0$ and $\dot{S}_1 < 0$, (1.5.66) yields the hyperbola with (without) limit points for $\epsilon > 0$ ($\epsilon < 0$). It, then, follows that if

$$\frac{2S'_{11}S_{221}}{S_{111}S'_{22}} > 1 \tag{1.5.71}$$

the parabola (1.5.70) is associated with the branch having limit points (Figure 1.32), and if

$$\frac{2S'_{11}S_{221}}{S_{111}S'_{22}} < 1 \tag{1.5.72}$$

(1.5.70) gives the bifurcation points on the other branch (Figure 1.33).

The inequality (1.5.71) must, of course, be the condition ensuring that the line (1.5.68) lies between the asymptotes of (1.5.66). To verify this, set $\epsilon = 0$ in (1.5.66) to obtain the asymptotes

$$u_1 = 0, \quad \lambda = -\frac{S_{111}}{2S'_{11}} u_1$$

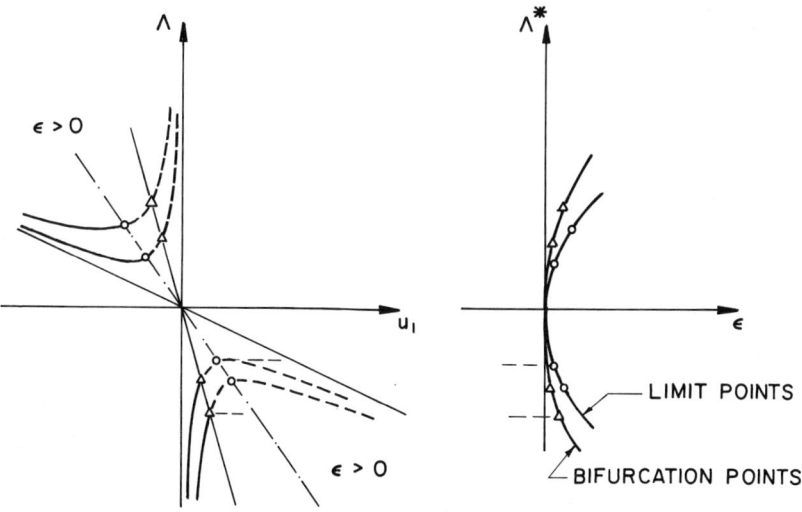

Figure 1.32

1.5 Coincident critical points

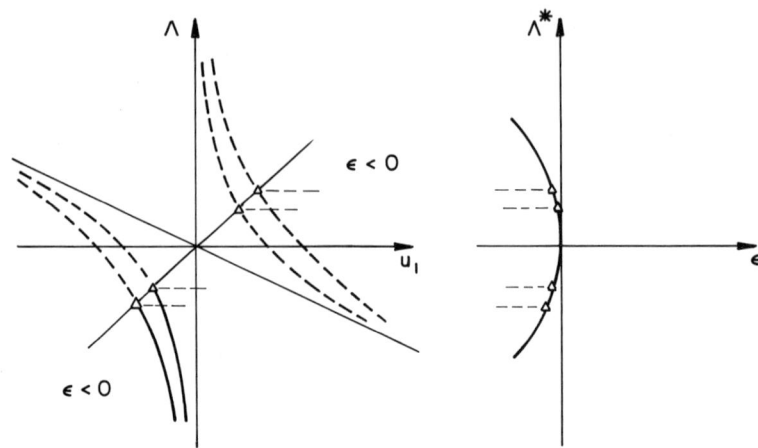

Figure 1.33

The line (1.5.68) remains between these asymptotes provided

$$-\frac{S_{221}}{S'_{22}} < -\frac{S_{111}}{2S'_{11}}$$

which leads to (1.5.71).

When (1.5.71) holds, one would naturally like to know whether the bifurcation occurs before or after the limit point. It is recalled that the critical line (dash-dotted in Figure 1.32) passing through the limit points is given by

$$\overset{*}{\lambda} = -\frac{S_{111}}{S'_{11}}\overset{*}{u}_1,$$

and the line (1.5.68) should, then, lie between this critical line and $u_1 = 0$ if the intersection is to occur before the limit point. That is,

$$-\frac{S_{221}}{S'_{22}} < -\frac{S_{111}}{S'_{11}}$$

or

$$\frac{S'_{11}S_{221}}{S_{111}S'_{22}} > 1 \tag{1.5.73}$$

It is not difficult to show that, if (1.5.73) holds the curvature of the parabola (1.5.70) is smaller than the curvature of (1.5.69), indicating higher imperfection-sensitivity. When the line (1.5.68) coincides with the critical line, then,

$$\frac{S_{221}}{S'_{22}} = \frac{S_{111}}{S'_{11}},$$

1.5.6 Illustrative example

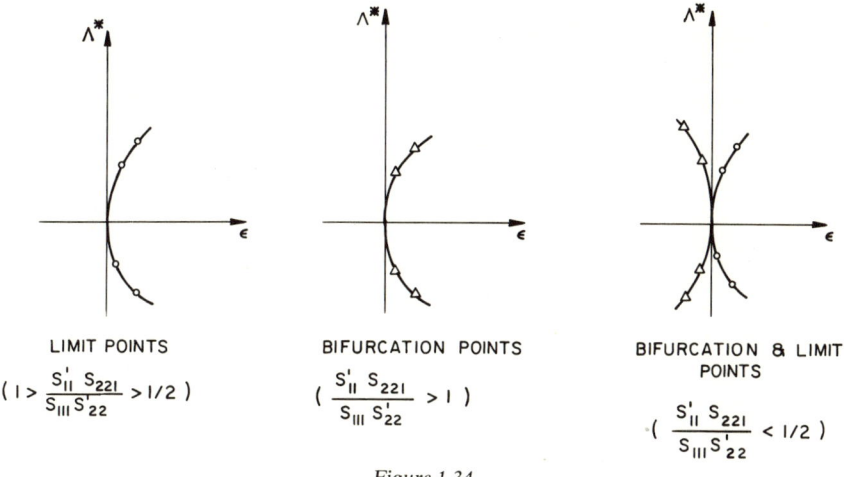

Figure 1.34

and (1.5.70) becomes identical to (1.5.69), bifurcation and limit points occurring simultaneously.

Figure 1.34 shows the dominating critical load-imperfection relationships under three distinct circumstances.

Problem. Discuss the behaviour of the system when the potential energy function is symmetric in u_2, and hence not only S_{211} but also S_{222} vanishes. [Hint: multiply the first equation in (1.5.67) by $S_{222} \neq 0$ and the second equation by $(S_{222})^2 \neq 0$, and then let S_{222} tend to zero].

1.5.6 Illustrative example

Consider the simple two-degree-of-freedom structural model presented by Chilver [70] and shown in Figure 1.35. It consists of three rigid links hinged at their joints and restrained by four elastic nonlinear springs as shown. The stability of this model under the axial force P will be studied.

The generalized coordinates q_1 and q_2 are chosen such that the extensions of the springs a, b, c and the rotation of d are given by

$$e_a = \tfrac{1}{2}L(q_1 + q_2), \quad e_b = \tfrac{1}{2}L(q_1 - q_2), \quad e_c = Lq_1$$

and (1.5.74)

$$\phi_d = \sin^{-1} 2q_2$$

respectively.

The strain energy stored in each of the springs is assumed to be in the

1.5 Coincident critical points

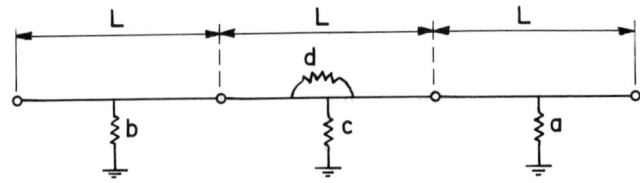

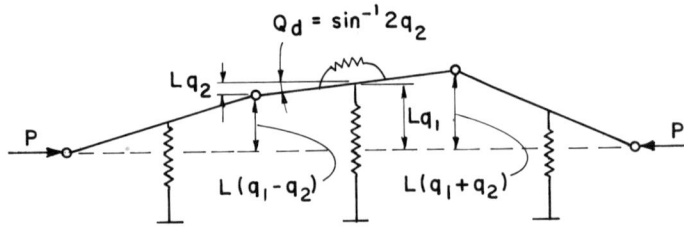

Figure 1.35

following forms:

$$U_a = \tfrac{1}{2}A(q_1+q_2)^2 + \tfrac{1}{6}B_1(q_1+q_2)^3 + \tfrac{1}{24}C(q_1+q_2)^4$$
$$U_b = \tfrac{1}{2}A(q_1-q_2)^2 + \tfrac{1}{6}B_2(q_1-q_2)^3 + \tfrac{1}{24}C(q_1-q_2)^4$$
$$U_c = \tfrac{1}{2}A_1 q_1^2 + \tfrac{1}{6}B q_1^3 + \tfrac{1}{24}C_1 q_1^4 \tag{1.5.75}$$
$$U_d = \tfrac{1}{2}\bar{A}_2(\sin^{-1} 2q_2)^2 \cong \tfrac{1}{2}\bar{A}_2(4q_2^2 + \tfrac{16}{3}q_2^4)$$
$$\equiv \tfrac{1}{2}A_2 q_2^2 + \frac{1}{4!}C_2 q_2^4$$

in which A's, B's and C's are constants.

Axial displacement of the force can be obtained from geometry as

$$E = L[q_1^2 + 3q_2^2 + \tfrac{1}{4}(q_1^4 + 6q_1^2 q_2^2 + 9q_2^4)]$$

in which terms up to the 4th order have been retained.

The total potential energy of the system is, then, given by

$$V = \tfrac{1}{2}(2A + A_1)q_1^2 + \tfrac{1}{2}(2A + A_2)q_2^2$$
$$+ \frac{1}{3!}[(B + B_1 + B_2)q_1^3 + 3(B_1 - B_2)q_1^2 q_2 + 3(B_1 + B_2)q_1 q_2^2$$
$$+ (B_1 - B_2)q_2^3] + \frac{1}{4!}[(2C + C_1)q_1^4 + 12C q_1^2 q_2^2$$
$$+ (2C + C_2)q_2^4] - PL[(q_1^2 + 3q_2^2) + \tfrac{1}{4}(q_1^4 + 6q_1^2 q_2^2 + q_2^4)] \tag{1.5.76}$$

1.5.6 Illustrative example

Evidently $q_1 = q_2 = 0$ describes the fundamental path along which the quadratic form of the energy is diagonalized. It then follows that

$$V(q_1, q_2, \Lambda) \equiv S(u_1, u_2, \Lambda) \tag{1.5.77}$$

in which the loading parameter is chosen as

$$\Lambda = PL/(A + \tfrac{1}{2}A_1) \tag{1.5.78}$$

The theory of Section 1.5.3 and/or 1.5.4 is, then, immediately applicable if a coincident critical point arises. Critical loads of the system are given by $S_{11}(0, 0, \Lambda) = 0$ and $S_{22}(0, 0, \Lambda) = 0$ which yield

$$\Lambda^c = 1$$
and $\tag{1.5.79}$
$$\Lambda^c = (2A + A_2)/3(2A + A_1)$$

respectively.

These critical loads are equal if

$$A_2 - 3A_1 = 4A \tag{1.5.80}$$

Assuming that (1.5.80) holds and there are no particular symmetry properties, the post-buckling paths are defined by (1.5.30) which in this case reads

$$\begin{aligned} S_{111} + 2S_{112}u_{2,1} + S_{122}(u_{2,1})^2 + 2S'_{11}\Lambda_1 &= 0 \\ S_{211} + 2S_{212}u_{2,1} + S_{222}(u_{2,1})^2 + 2S'_{22}\Lambda_1 u_{2,1} &= 0 \end{aligned} \tag{1.5.81}$$

where u_1 has been given the role of η. The energy coefficients evaluated at the critical point are

$$\begin{aligned} S_{111} &= B_1 + B_2 + B, & S_{112} &= B_1 - B_2, \\ S_{122} &= B_1 + B_2, & S_{222} &= B_1 - B_2 \\ S'_{11} &= -(2A + A_1), & S'_{22} &= -3(2A + A_1) \\ S_{1111} &= 2C + C_1 - 3(2A + A_1) \\ S_{1112} &= S_{1222} = 0, & S_{1122} &= 2C - 3(2A + A_1) \\ S_{2222} &= 2C + C_2 - 27(2A + A_1) \end{aligned} \tag{1.5.82}$$

Two particular cases will be considered:
(i) Suppose $B = 0$ and $B_2/B_1 = -45/7$; then, upon substituting for the coefficients into (1.5.81) one has the three solutions

$$u_{2,1} = 2; \quad u_{2,1} = -0.357; \quad u_{2,1} = 0.638 \tag{1.5.83}$$

Assuming further that $\dfrac{2A + A_1}{B_1} = \dfrac{1}{2}$, and substituting for $u_{2,1}$ into (1.5.81)

1.5 Coincident critical points

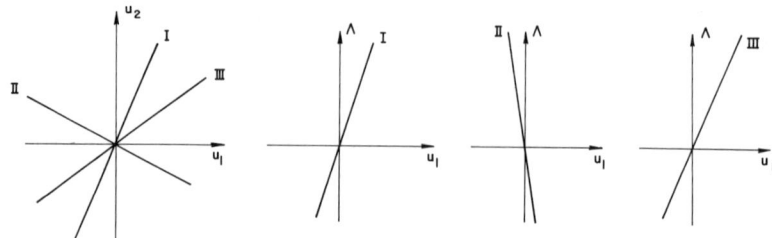

Figure 1.36

yields

$$\Lambda_1 = 2 \cdot 57; \quad \Lambda_1 = -11 \cdot 41; \quad \Lambda_1 = 1 \cdot 83 \qquad (1.5.84)$$

respectively. These solution rays are shown in Figure 1.36.
(ii) Consider now the case in which $B_2/B_1 = 25/63$ and $B = 0$; equations (1.5.81), then, yield one real solution only,

$$u_{2,1} = 0 \cdot 25, \qquad (1.5.85)$$

and if it is further assumed that $(2A + A_1)/B_1 = \frac{1}{2}$, one obtains

$$\Lambda_1 = 1 \cdot 79 \qquad (1.5.86)$$

Next, consider symmetry in the form

$$S(\Lambda, u_1, u_2) = S(\Lambda, u_1, -u_2)$$

which yields

$$S_{112} = S_{222} = 0$$

so that

$$B_1 = B_2 \qquad (1.5.87)$$

and hence

$$S_{111} = 2B_1 + B, \qquad S_{122} = 2B_1 \qquad (1.5.88)$$

(1.5.81) takes then the form

$$\begin{aligned} 2B_1 + B + 2B_1(u_{2,1})^2 + 2S'_{11}\Lambda &= 0 \\ 4B_1 u_{2,1} + 2S'_{22}\Lambda_1 u_{2,1} &= 0 \end{aligned} \qquad (1.5.89)$$

which results in the three solution rays

$$u_{2,1} = 0, \quad \Lambda_1 = -(2B_1 + B)/2S'_{11}$$

and

$$\Lambda_1 = -2B_1/S'_{22}, \quad u_{2,1} = \pm\sqrt{-\left(\frac{B}{2B_1} + \frac{1}{3}\right)}$$

1.5.6 Illustrative example

Note that these solutions can also be obtained directly from (1.5.40) if it is recognized that $u_2 \equiv z_1$ and $u_1 \equiv u_2$. Figure 1.37 shows the general forms of the solutions.

Finally, consider symmetry in both u_1 and u_2 implying

$$S_{111} = S_{112} = S_{122} = S_{222} = 0$$

which leads to

$$B_1 = B_2 = B = 0$$

and the identities $u_1 \equiv z_1$, $u_2 \equiv z_2$ as in the theory.

Evidently, the post-buckling paths are now given by (1.5.41) and (1.5.43) in the theory. One notes first that due to *individual* symmetry,

$$z_{2,1} = 0 \quad \text{and} \quad z_{1,2} = 0$$

define two uncoupled paths in the z_1 and z_2 modes respectively.

Having established the existence of the path $z_{1,2} = 0$ in the z_2 direction, z_1 can be taken as the perturbation parameter η. The equation (1.5.43), then, takes the form

$$S_{1111} + 3S_{1122}(z_{2,1})^2 + 3S'_{11}\Lambda_{11} = 0 \quad \text{for} \quad \alpha = 1$$

and (1.5.90)

$$S_{2222}(z_{2,1})^3 + 3S_{2211}z_{2,1} + 3S'_{22}\Lambda_{11}z_{2,1} = 0 \quad \text{for} \quad \alpha = 2$$

which yields

$$z_{2,1} = 0$$

as one of the solutions, confirming the earlier result. Using the coefficients (1.5.82) and $z_{2,1} = 0$, and solving (1.5.90) simultaneously result in two more derivatives,

$$z_{2,1} = \pm\sqrt{\frac{-3C_1}{16C - C_2}} \qquad (1.5.91)$$

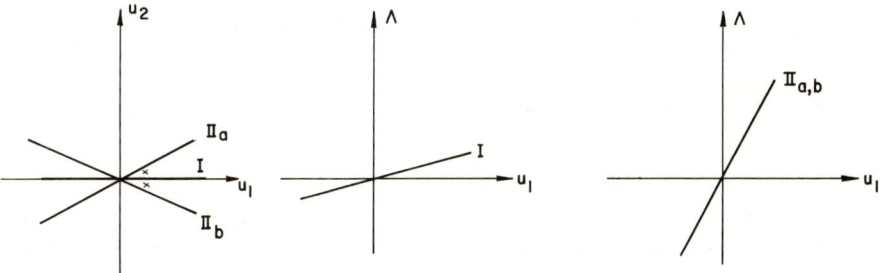

Figure 1.37

1.5 Coincident critical points

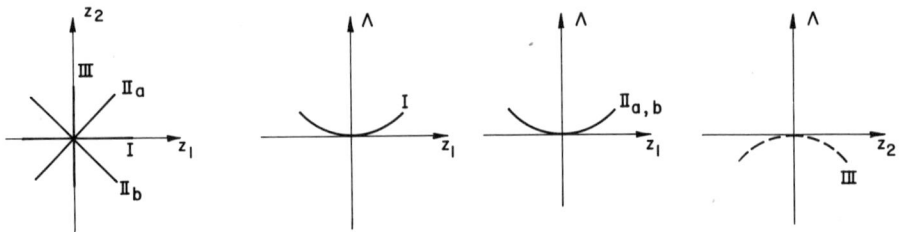

Figure 1.38

(1.5.41) ensures zero slopes Λ_1 (or Λ_2) and the corresponding curvatures, Λ_{11}, can be obtained from (1.5.90) as

$$\Lambda_{11} = \frac{2C + C_1}{3(2A + A_1)} - 1 \quad \text{for} \quad z_{2,1} = 0$$

and (1.5.92)

$$\Lambda_{11} = \frac{1}{3(2A + A_1)}\{2C + C_1 - 3(2A + A_1) - 9\frac{C_1}{16C - C_2}[2C - 3(2A + A_1)]\}$$

for (1.5.91).

The path $z_{1,2}$ was predicted earlier and it follows that a total of 4 paths may exist. If it turns out that the quantity under the square root in (1.5.91) is negative, then, there are only two paths along the axes z_1 and z_2.

Finally, the curvature Λ_{22} for the case $z_{1,2} = 0$ can readily be obtained by setting $\eta \equiv z_2$ in (1.5.46) as

$$\Lambda_{22} = -\frac{S_{2222}}{3S'_{22}} = \frac{2C + C_2}{9(2A + A_1)} - 3 \tag{1.5.93}$$

It is understood that, depending on the specific values assigned to A's and C's, the curvatures (1.5.92) and (1.5.93) can be positive or negative, and Figure 1.38 illustrates this fact schematically.

Problem 1. Following the theory in Section 1.5.2, investigate the stability of the coincident critical point arising in the analysis of the above model when the energy function is symmetric in u_2 only. Similarly, examine the stability of the coincident critical point when the energy function is symmetric in both u_1 and u_2, and demonstrate the relationship between the stability of this point and the curvatures (1.5.92) and (1.5.93).

Part 2

Multiple-parameter systems

2.1

Fundamentals

2.1.1 General remarks

In the development of the general theory of elastic stability it has been usually assumed that the external loading of a structure could be represented adequately by a single variable parameter. It is clear, however, that in practical problems more than one independent parameter may often be present. Thin rectangular plates under the combined action of axial compression and shear (or axial compression in both directions), cylindrical and conical shells under external pressure and axial compression, shallow arches and spherical caps under uniform pressure and concentrated load at the apex, and several particular problems concerned with frames and columns can be mentioned as examples.

It can even be argued that under most circumstances structural systems contain more than one independent parameter; if the weight of a structure is to be accounted for, for instance, the system becomes at least a two-parameter one. External loads or weight are not, of course, the only variables of interest in a structural problem; the effect of imperfections, certain magnitudes or a modulus on the buckling behaviour, for example, are often sought, and in all such cases an essentially multiple-parameter system arises.

Papkovich [80–81], considering the linear bifurcation buckling of a structural system with independent loading parameters, showed that the associated stability boundary (see Section 2.3.4) cannot have convexity towards the origin. A similar result was obtained by Buckens [82], who studied linear eigenvalue problems. Reference [83] contains the result as a special case. In these studies, it is assumed that the total potential energy of the system is a quadratic form, and the scope of the results, therefore, is limited to systems which exhibit a purely trivial fundamental solution involving no pre-buckling deflections.

A general nonlinear theory of elastic stability concerning the multiple-parameter systems has only recently been developed [84, 43 to

2.1 Fundamentals

53] and will be presented here. The theory contains most of the material discussed in Part 1 as a special case, and with appropriate interpretation of the parameters the results of Part 1 are immediately recoverable. Part 1 and Part 2, however, have been treated separately and essentially in an independent manner in order to increase the accessibility of the material in Part 1.

It has been observed [43, 84] that the two well-known critical points, namely bifurcation and limit points, cannot describe the instability behaviour of multiple-parameter systems adequately, and a reclassification of the critical conditions characterizing the buckling behaviour more aptly has been presented.

Chapter 2.1 is devoted to introducing the basic principles, concepts, definitions and a preliminary discussion leading to such a reclassification of critical points upon which the more detailed analyses of the subsequent chapters are based.

The discussion of multiple-parameter systems must not be regarded as one which is solely concerned with systems having several independent loading parameters. On the contrary, several significant aspects of the buckling and post-buckling behaviour of one-parameter systems and the circumstances giving rise to such behaviour are revealed and our understanding of certain concepts is facilitated through the perspective provided by a multiple-parameter analysis.

2.1.2 Basic concepts and definitions

A general elastic conservative structural system described by N generalized coordinates Q_i and M independent loading parameters Λ^j is considered. A given set of the Q_i ($i = 1, 2, \ldots, N$) defines the configuration of the system completely. The Λ^j ($j = 1, 2, \ldots, M$) may represent certain parameters of the system independent of the Q_i and of each other, and are not confined to describe external loads only. This freedom in the choice of parameters gives the general theory a desirable flexibility and provides for its direct application to a wide class of practical problems. Some results, however, are valid for external loads only, and appropriate restrictions will be imposed on the parameters to this effect in the related sections. Nevertheless, for convenience, these parameters will often be referred to as loads in the sequel.

Reference will often be made to two multi-dimensional spaces (*manifolds*), namely the M dimensional *load-space* and the $M+N$ dimensional *load-deflection space* (or *configuration space*), the former being a sub-space of the latter. It is assumed that there is a one-to-one

2.1.2 Basic concepts and definitions

correspondence between a set of variables (Q_i, Λ^j) and the points of the load-deflection space.

An equilibrium state corresponding to a given set of Λ^j is associated with a point in the $M+N$ dimensional load-deflection space, and the entirety of these equilibrium points constitutes an M dimensional *equilibrium surface* (*hyper-surface*). It must immediately be remarked that, although a one-to-one correspondence between the points of the load-deflection space and a specific set of $M+N$ variables (Q_i, Λ^j) is assumed to exist, it does not necessarily follow that there is also a one-to-one correspondence between the sets of Λ^j and equilibrium states of the system. In fact, the points of the load-space will, in general, be associated with more than one equilibrium state which can be stable or unstable, stability being defined precisely as in Part 1. Some of the points of the load-space may correspond to states of critical equilibrium, and the entirety of these points will be called *the critical surface*. Both the equilibrium surface and the critical surface may consist of several self-curvilinear surfaces, and considering all possible sequences of loading from the unloaded state, some critical surfaces will be associated with an *initial* loss of stability constituting the so-called *stability boundary* of the system. In other words, the latter is a specific part of the former. It will be seen later (Section 2.2.4) that, under some circumstances, there may be no neighbouring equilibrium states beyond the stability boundary, and then, the stability boundary can more aptly be termed *the existence boundary* demarcating *the regions of existence* and *inexistence*.

It is understood that the critical surfaces and consequently the stability boundary are defined in the load-space and not in the load-deflection space. In fact, they are the projections of the *critical zone* (consisting of critical equilibrium states) of the equilibrium surface into the load-space. This is illustrated schematically on a strip of the equilibrium surface in Figure 2.1.

Having introduced some of the basic concepts of the theory, the total potential energy of the system will now be brought into the picture which will form the basis of analytical definitions and the analyses in the sequel. For each particular set of values assigned to the Λ^j, the system is assumed to have a total potential energy consisting of the strain energy and potential energies of applied loads. One can, then, define a total potential energy function

$$V = V(Q_i, \Lambda^j) \qquad (2.1.1)$$

which depends on $N+M$ independent variables Q_i and Λ^j. It is further assumed that this function is single-valued and continuously differentiable at least in the region of interest.

2.1 Fundamentals

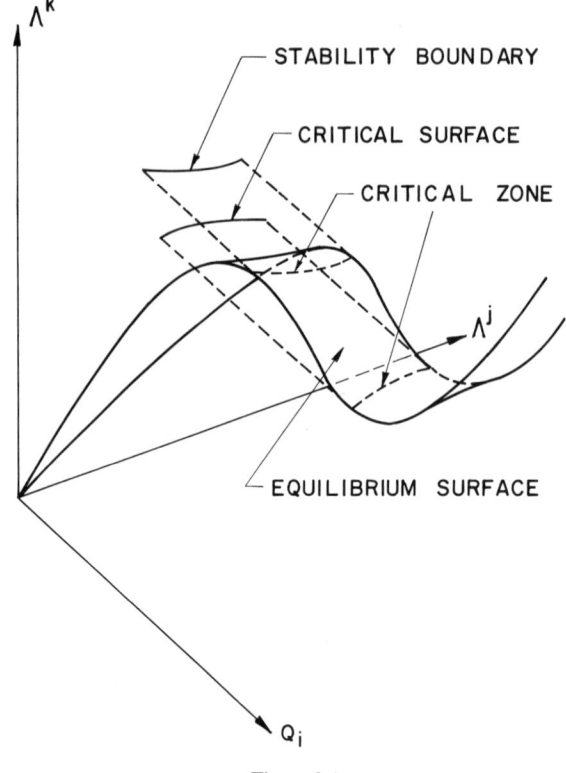

Figure 2.1

In many structural problems, and in particular when the Λ^j represent external loads only, the function (2.1.1) can be expressed as

$$V = U(Q_i) - \Lambda^j E^j(Q_i) \tag{2.1.2}$$

where U is the strain energy and E^j are the generalized displacements corresponding to Λ^j, and the summation convention applies. This specialized system, being linear in the Λ^j, may exhibit certain additional characteristics which will be explored in appropriate sections.

The equilibrium equations

$$V_i(Q_j, \Lambda^k) = 0, \quad i = 1, 2, \ldots, N, \tag{2.1.3}$$

evidently, define an M dimensional equilibrium surface in the $M+N$ dimensional load-deflection space. To judge by what has been learned from the equilibrium paths in Part 1, it is expected that there is a close connection between the shape of the equilibrium surface and the buckling

2.1.2 Basic concepts and definitions

behaviour of the system. One would, therefore, aim at examining this surface as thoroughly as possible.

Following the same line of reasoning as in Part 1, the overall features of the equilibrium surface will initially be explored through an analysis based on Taylor's expansion as a prelude to more detailed investigation of specific aspects via the multiple-parameter perturbation technique which will be facilitated with the introduction of appropriate transformations.

Thus, suppose the equilibrium equations (2.1.3) can be solved simultaneously to yield the equilibrium surface in the form

$$Q_i = Q_i(\Lambda^j) \quad \begin{aligned} i &= 1, 2, \ldots, N; \\ j &= 1, 2, \ldots, M, \end{aligned} \quad (2.1.4)$$

and consider an arbitrary point F on it representing a state of equilibrium $Q_i^F(\Lambda_F^j)$ which will be called fundamental. Using the q_i and λ^j to denote increments in the Q_i and Λ^j respectively, the potential energy of the system is referred to the fundamental state, F, by writing it in the form

$$V = V(Q_i^F + q_i, \Lambda_F^j + \lambda^j) \quad (2.1.5)$$

Introduce now the linear orthogonal transformation

$$q_i = \alpha_{ij} u_j, \quad \alpha_{ij}\alpha_{jk} = \delta_{ik} \quad (2.1.6)$$

to diagonalize the quadratic form (in the q_i) of the energy expansion around the fundamental state. Substituting for the q_i into (2.1.5) one has, in analogy with Section 1.1.2,

$$H(u_i, \lambda^j) \equiv V(Q_i^F + \alpha_{ij}u_j, \Lambda_F^j + \lambda^j) \quad (2.1.7)$$

where the u_i are again the principal generalized coordinates (Figure 2.2).

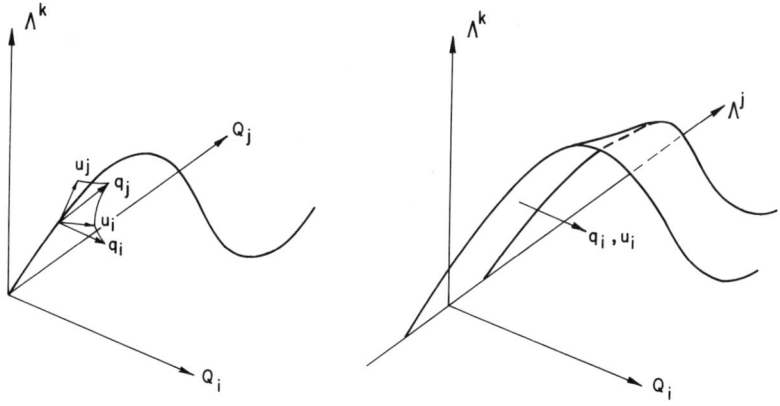

Figure 2.2

2.1 Fundamentals

Taylor's expansion yields

$$H = H_0 + H_i u_i + H^i \lambda^i$$

$$+ \frac{1}{2!}(H_{ii}u_i^2 + 2H_i^i \lambda^i u_i + H^{ij}\lambda^i \lambda^j)$$

$$+ \frac{1}{3!}(H_{ijk}u_i u_j u_k + 3H_{ij}^k u_i u_j \lambda^k + 3H_i^{jk} u_i \lambda^j \lambda^k + H^{ijk}\lambda^i \lambda^j \lambda^k)$$

$$+ \frac{1}{4!}(\cdots) + \frac{1}{5!}(\cdots) + \cdots \qquad (2.1.8)$$

Here, and in the sequel, the summation convention is adopted both for subscripts and superscripts separately. It is also understood that upper and lower indices on the H coefficients denote partial differentiation with respect to the corresponding loading parameter and principal generalized coordinate respectively, all derivatives being evaluated at the fundamental state of equilibrium F.

It is noted that the coefficients H_i in the first line are all zero since F is a point of equilibrium, and the equilibrium equations can be obtained by differentiating (2.1.8) with respect to the u_i, e.g.,

$$\frac{\partial H}{\partial u_r} = H_{rr} u_r + H_r^i \lambda^i$$

$$+ \frac{1}{2!}(H_{rij} u_i u_j + 2H_{ri}^j u_i \lambda^j + H_r^{ij} \lambda^i \lambda^j)$$

$$+ \frac{1}{3!}(H_{rijk} u_i u_j u_k + 3H_{rij}^k u_i u_j \lambda^k + 3H_{ri}^{jk} u_i \lambda^j \lambda^k + H_r^{ijk} \lambda^i \lambda^j \lambda^k$$

$$+ \frac{1}{4!}(\cdots) + \cdots = 0 \qquad (2.1.9)$$

2.1.3 Classification of critical points

The equilibrium surface can now be examined locally. In general, all the coefficients are finite, and the equilibrium equations $\partial H / \partial u_i = 0$ ($i = 1, 2, \ldots, N$) yield to a first approximation

$$H_{ii} u_i + H_i^j \lambda^j = 0 \qquad (2.1.10)$$

These equations determine an M dimensional plane and indicate a one-to-one correspondence between a set of Λ^j and a set of u_i. The plane can be envisaged as a small local part of the equilibrium surface around the point F, and it is concluded that there is a *unique* equilibrium surface through a noncritical equilibrium state. This is the most general behaviour associated with the equilibrium of the system.

2.1.3 Classification of critical points

The stability or instability of an equilibrium state here is decided in exactly the same way as discussed in Sections 1.1.1 and 1.1.4 (see Table 2), and the discussion will not, therefore, be repeated. Attention throughout Part 2, will be focussed on simple (discrete) critical points at which only one of the stability coefficients, H_{ii}, vanishes.

Suppose F is moved on the equilibrium surface to a simple critical state of equilibrium at which, say $H_{11} = 0$ and $H_{ss} \neq 0$ (for all $s \neq 1$). Now the most important factor which appears to determine the shape of the equilibrium surface and the type of the critical point is

$$\overrightarrow{\text{grad}}_\lambda H_1(0, 0).$$

Two distinct cases arise and they will be studied separately:

(i) **The case in which** $\overrightarrow{\text{grad}}_\lambda H_1(0, 0) \neq 0$

This condition implies that at least one of the coefficients $H_1^i(0, 0)$ does not vanish. Such a critical point will be termed *general*, and it is understood that in practical situations one is interested in the *primary general* point at which all $H_{ss} > 0$ and the stability is lost initially.

Supposing for now that all the remaining coefficients are nonzero, the equilibrium equations can be written as

$$H_1^i \lambda^i + \tfrac{1}{2} H_{1ij} u_i u_j + \cdots = 0 \tag{2.1.11}$$

and

$$H_{ss} u_s + H_s^i \lambda^i + \tfrac{1}{2} H_{s11} u_1^2 + \cdots = 0 \tag{2.1.12}$$

which define an M dimensional equilibrium surface in the $M + N$ dimensional load-deflection space. Substituting for u_s into (2.1.11) from (2.1.12) and keeping to a first approximation yields

$$H_1^i \lambda^i + \tfrac{1}{2} H_{111} u_1^2 = 0 \tag{2.1.13}$$

and substituting for u_1 into (2.1.12) results in

$$H_{ss} u_s + \left(H_s^i - \frac{H_1^i H_{s11}}{H_{111}} \right) \lambda^i = 0 \tag{2.1.14}$$

Equations (2.1.13) and (2.1.14) can be regarded as the projections of the equilibrium surface into the $u_1 - \lambda^i$ and $u_s - \lambda^i$ subspaces respectively, the former representing a curved surface and the latter a plane. In other words the critical coordinate ceases to be single-valued while the noncritical coordinates remain as linear functions of the λ^i as in the case of noncritical equilibrium points. These projections are shown schematically in Figures 2.3 and 2.4.

2.1 Fundamentals

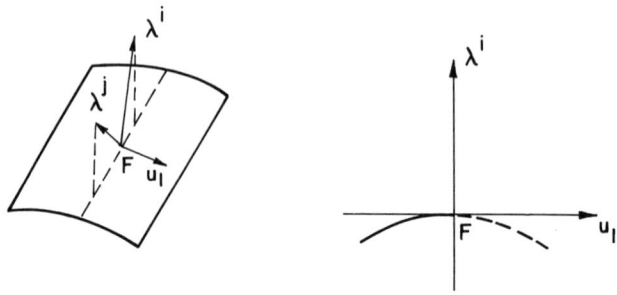

Figure 2.3

It is interesting to note that the equations (2.1.13) and (2.1.14) resemble the corresponding equations (1.1.23) of a limit point in Section 1.1.2, the only difference being the summation on the λ^i. In fact, the former equations reduce to the latter if one assumes that the λ^i are functions of a single variable loading parameter ξ. Expanding the functions $\lambda^i(\xi)$ around the point F where $\lambda^i = \xi = 0$ one gets

$$\lambda^i = l^i \xi + \tfrac{1}{2} k^i \xi^2 + \cdots \quad (2.1.15)$$

in which l^i, k^i, ... are constants.

Substituting for the λ^i into (2.1.13) yields the first order result

$$H^i_1 l^i \xi + \tfrac{1}{2} H_{111} u_1^2 = 0 \quad (2.1.16)$$

in which $H^i_1 l^i$ is immediately recognized as

$H'_1 (\equiv \partial H_1 / \partial \xi |_{\xi=0})$ giving finally

$$H'_1 \xi + \tfrac{1}{2} H_{111} u_1^2 = 0$$

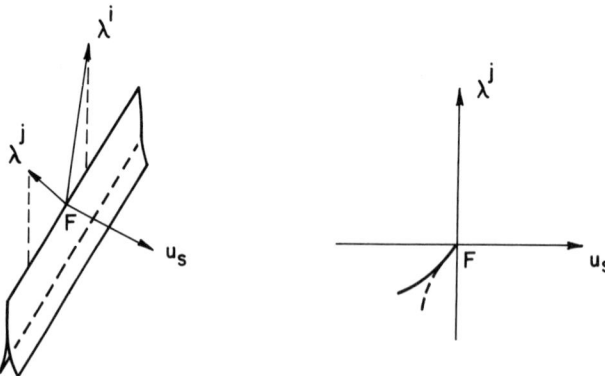

Figure 2.4

2.1.3 Classification of critical points

which is, indeed, the first relationship in (1.1.23) (Figure 2.5b). Similarly, (2.1.14), reduces to the second equation in (1.1.23).

An important distinction between the systems with a single loading parameter and those with several parameters can now be drawn; thus, in contrast with the former, the conditions $H_1^i \neq 0$ ($i = 1, 2, \ldots, N$) do not ensure a limit point in the latter since the linear form $H_1^i l^i$ might, then, vanish depending on the l^i (i.e., on the loading) in which case the equilibrium equations must be reconsidered and additional terms retained. If this is done, one obtains to a first approximation

$$\tfrac{1}{2}H_{111}u_1^2 + au_1\xi + \tfrac{1}{2}b\xi^2 = 0 \tag{2.1.17}$$

where $a = \left(H_{11}^i - H_{s11}\dfrac{H_s^i}{H_{ss}}\right)l^i \equiv c^i l^i$

$$b = \left(H_1^{ij} + H_{1sr}\dfrac{H_s^i H_r^j}{H_{ss}H_{rr}} - 2H_{1s}^i\dfrac{H_s^j}{H_{ss}}\right)l^i l^j + H_1^i k^i$$

$$\equiv d^{ij}l^i l^j + H_1^i k^i \equiv d + H_1^i k^i \tag{2.1.18}$$

and

$$H_{ss}u_s + H_s^i l^i \xi = 0 \tag{2.1.19}$$

Solving (2.1.17) for u_1 yields

$$u_1 = \dfrac{1}{H_{111}}[-a \pm (a^2 - H_{111}b)^{\tfrac{1}{2}}]\xi \tag{2.1.20}$$

which indicates bifurcation on a plot of u_1 versus ξ (Figure 2.5a) provided

$$a^2 - H_{111}b > 0 \tag{2.1.21}$$

This phenomenon can be seen in another way; suppose some of the coefficients H_1^i vanish (say $H_1^t = 0$, $t = 1, \ldots, M_1$; $M_1 < M$), then, the equilibrium equations take the form

$$H_1^x\lambda^x + \tfrac{1}{2}H_{1ij}u_i u_j + H_{1j}^t u_j \lambda^t + \tfrac{1}{2}H_1^{rt}\lambda^r\lambda^t + \cdots = 0 \tag{2.1.22}$$

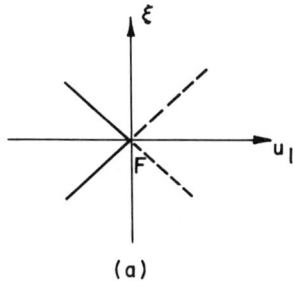

Figure 2.5

2.1 Fundamentals

and

$$H_{ss}u_s + H_s^i\lambda^i + \tfrac{1}{2}H_{s11}u_1^2 + \cdots = 0 \tag{2.1.23}$$

where $s \neq 1$, t and r range from 1 to M_1, and x ranges from $M_1 + 1$ to M (i.e., $x \neq t$, $x \neq r$), summations being carried out on repeated subscripts and superscripts, as before, over admissible ranges.

Substituting for u_s in the equation (2.1.22) and keeping to a first order approximation yields

$$H_1^x\lambda^x + \tfrac{1}{2}H_{111}u_1^2 + c^t u_1\lambda^t + \tfrac{1}{2}d^{tr}\lambda^t\lambda^p = 0 \tag{2.1.24}$$

and substituting for u_1 into (2.1.23) results in

$$H_{ss}u_s + \left(H_s^x - \frac{H_1^x H_{s11}}{H_{111}}\right)\lambda^x + H_s^t\lambda^t = 0 \tag{2.1.25}$$

where $s \neq 1$, $x \neq t$, $x \neq p$, and the coefficients c^t and d^{tr} are given by (2.1.18) (note, however, that the summations now range from 1 to M_1, $M_1 < M$).

It is now clear that bifurcation is not ruled out; in fact, if the loading follows the ray

$$\begin{aligned}\lambda^x &= 0 \\ \lambda^t &= l^t\xi, \quad t \neq x \end{aligned} \tag{2.1.26}$$

the equilibrium equation (2.1.24) yields

$$\tfrac{1}{2}H_{111}u_1^2 + au_1\xi + \tfrac{1}{2}b\xi^2 = 0 \tag{2.1.27}$$

in which

$$a = c^t l^t$$
$$b = d^{tr}l^t l^r$$

Solution of (2.1.27) for u_1 indicates bifurcation as illustrated in Figure 2.5a, provided

$$a^2 - H_{111}b > 0$$

On the other hand, suppose a more general ray, such that all $\lambda^i \neq 0$ or at least some $\lambda^x \neq 0$ is followed; equation (2.1.24) then yields

$$H_1^x l^x \xi + \tfrac{1}{2}H_{111}u_1^2 = 0 \tag{2.1.28}$$

provided $H_1^x l^x \neq 0$, indicating a limit point. In the event that $H_1^x l^x = 0$, one returns to the case considered before.

Finally, consider the special situation in which an additional coefficient, namely, H_{111} vanishes. This may happen due to certain symmetry properties, and the first order equilibrium equation corresponding to

2.1.3 Classification of critical points

(2.1.24), is then, given by

$$H_1^x \lambda^x + \frac{1}{3!} \bar{H}_{1111} u_1^3 + c^t u_1 \lambda^t + \tfrac{1}{2} d^{tr} \lambda^t \lambda^r = 0 \qquad (2.1.29)$$

where $\bar{H}_{1111} = H_{1111} - 3 \sum_{s=2}^{N} \dfrac{(H_{s11})^2}{H_{ss}}$.

Any ray involving λ^x now leads to

$$H_1^x l^x \xi + \frac{1}{3!} \bar{H}_{1111} u_1^3 = 0 \qquad (2.1.30)$$

which defines a third order curve having a point of inflexion at F and being stable (if all $H_{ss} > 0$), except at F which may be unstable. On the other hand, a ray which is in the λ^t subspace leads to

$$\frac{1}{3!} \bar{H}_{1111} u_1^3 + a u_1 \xi + \tfrac{1}{2} b \xi^2 = 0 \qquad (2.1.31)$$

If this equation is differentiated with respect to u_1 twice and evaluated at F, one has

$$d\xi/du_1 |_F = 0 \quad \text{or} \quad d\xi/du_1 |_F = -2a/b$$

which means a symmetric point of bifurcation, and it will be seen later that such a point takes a singular position in the overall instability picture resulting in an abrupt change in the shape of the stability boundary. The general critical point will, then, be called *singular* indicating this phenomenon.

One concludes that a general critical point can be considered as a limit point and/or bifurcation point depending on the mode of loading. The distinguishing feature of a general critical point is that the equilibrium surface in its vicinity is a *proper* continuous surface while this is not so in the case of $\overrightarrow{\text{grad}}_\lambda H_1(0, 0) = 0$ to be studied next.

(ii) **The case in which** $\overrightarrow{\text{grad}}_\lambda H_1(0, 0) = 0$

This condition implies that all the coefficients H_1^i ($i = 1, 2, \ldots, N$) vanish at the critical point F where $H_{11} = 0$, $H_{ss} \neq 0$ for $s \neq 1$. This is, evidently, a special critical state compared to what has been discussed in case (i), and it will be called *special*.

Assuming that all the other coefficients are nonzero, the first order solution of the equilibrium equations takes the form

$$\tfrac{1}{2} H_{111} u_1^2 + c^i u_1 \lambda^i + \tfrac{1}{2} d^{ij} \lambda^i \lambda^j = 0 \qquad (2.1.32)$$

and

$$H_{ss} u_s + H_s^i \lambda^i = 0.$$

119

2.1 Fundamentals

An arbitrary ray $\lambda^i = l^i \xi$, where some $l^i \neq 0$, leads to

$$u_1 = \frac{1}{H_{111}}[-a \pm (a^2 - H_{111}b)^{\frac{1}{2}}]\xi \qquad (2.1.33)$$

indicating bifurcation provided $a^2 - H_{111}b > 0$.

It is noted that limit points are definitely ruled out, i.e., there exists no rays with respect to which the critical point F can be considered a limit point. Thus, it is concluded that a sufficient condition of bifurcation buckling is given by

$$\overrightarrow{\mathrm{grad}}_\lambda H_1(0,0) = 0$$

with the side condition

$$a^2 - H_{111}b > 0. \qquad (2.1.34)$$

The surface (2.1.32) is now an *improper* (*degenerate*) one, and for $M = 2$, for example, it is a *degenerate quadric*. In the majority of structural problems it consists of two intersecting planes. This can be seen more clearly if it is further assumed that all $H_1^i = 0$, and the energy function V is linear in the Λ^i as in the equation (2.1.2). The latter assumption implies

$$V_i^{jk} = V_i^{jkl} = \cdots = V_{ij}^{kl} = \cdots = 0$$

which, in turn, and on the basis of the transformation (2.1.6) leads to

$$H_i^{jk} = H_i^{jkl} = \cdots = H_{ij}^{kl} = \cdots = 0 \qquad (2.1.35)$$

If these assumptions are introduced into the energy function, the equilibrium equation (2.1.32) reduces to

$$u_1(\tfrac{1}{2}H_{111}u_1 + H_{11}^i \lambda^i) = 0, \qquad (2.1.36)$$

and noncritical coordinates are expressed as

$$H_{ss}u_s + \tfrac{1}{2}H_{s11}u_1^2 = 0.$$

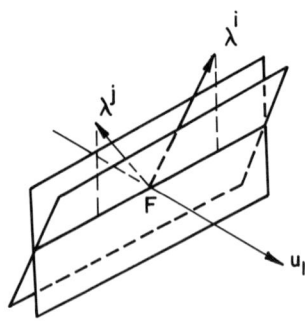
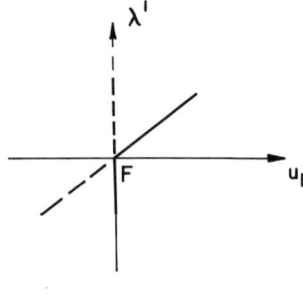

Figure 2.6

2.1.3 Classification of critical points

Equation (2.1.36), clearly represents two intersecting multidimensional planes (Figure 2.6), $u_1 = 0$ defining a fundamental plane and the expression in the parentheses a post-buckling plane in the vicinity of F.

It has, thus, been demonstrated analytically that the buckling behaviour of structures under combined loading can normally be described either by a *general* or a *special* critical point on the equilibrium surface. The former is often associated with a limit point, but under some conditions it appears as a point of bifurcation while at the latter only bifurcation can occur, a simple extremum being ruled out. The equilibrium surface in the vicinity of a general critical point is normally a *proper* surface, borrowing the term from the three dimensional spaces. On the other hand, in the vicinity of a special critical point, the equilibrium surface is an *improper* or *degenerate* one, often involving the intersection of two multi-dimensional planes.

2.2

General critical points

In this and the following chapters, the basic principles and concepts introduced in 2.1 will further be developed via the systematic multiple-parameter perturbation technique described in Section 1.1.3. Initial post-critical characteristics of multiple-parameter systems will thus be discussed with reference to *general* and *special* critical points separately in greater detail and with particular emphasis on the associated stability boundary. The determination of a stability boundary arises as a significant problem in connection with combined loading, and often poses fundamental difficulties, particularly when the systems develop large pre-buckling deflections. In such situations, lower and/or upper bound estimates of the stability boundary can be helpful and are often sought. In an effort to provide a basis for such estimates several theorems concerning the basic properties of the stability boundary with regard to certain well-defined classes of systems have been formulated [43 to 53], and will be presented in appropriate sections. These theorems, establishing a relationship between the type of the critical point and the shape of the stability boundary, supply definite information about the *convexity* of the stability boundary which enables the analyst to obtain a lower or upper bound to it in many particular problems. A second approach is concerned with a more direct estimation of the stability boundary through successive perturbations similar to that applied in Section 1.1.5 for critical loads, and will be discussed in the following section. Although this is being done under 'general critical points', the method is valid for special as well as general critical points, again in analogy with Section 1.1.5.

2.2.1 Estimation of the stability boundary

Consider the structural systems characterized by the total potential energy function (2.1.1). The equilibrium surface defined by

$$V_i(Q_j, \Lambda^k) = 0 \qquad (2.2.1)$$

2.2 General critical points

in the load-deflection space may have various disconnected branches, and the attention now will be focussed on the one which passes through the origin. This surface will be called '*fundamental*'.

Suppose equations (2.2.1) can be solved simultaneously to yield the fundamental equilibrium surface in the parametric form

$$Q_i = Q_i(\eta^k) \qquad i = 1, 2, \ldots, N;$$
$$\Lambda^j = \Lambda^j(\eta^k) \qquad j, k = 1, 2, \ldots, M \qquad (2.2.2)$$

where the unspecified independent perturbation parameters η^k ($k = 1, 2, \ldots, M$) are chosen such that the functions (2.2.2) are single-valued.

Substituting the assumed solution (2.2.2) back into the equilibrium equations results in the identities

$$V_i[Q_j(\eta^l), \Lambda^k(\eta^l)] \equiv 0 \qquad (2.2.3)$$

which, when differentiated with respect to η^l ($l = 1, 2, \ldots, M$) successively, yield sets of ordered equations.

Thus, differentiating with respect to η^m ($m = 1, 2, \ldots, M$) and evaluating at the origin, one gets

$$V_{ij}Q_j^m + V_i^j \Lambda^{j,m} = 0 \qquad (2.2.4)$$

in which differentiation of the Q_i and Λ^j is indicated by superscripts, a comma being used in the latter. The second differentiation of (2.2.3) with respect to η^n ($n = 1, 2, \ldots, M$) yields, after evaluation at the origin,

$$(V_{ijk}Q_k^n + V_{ij}^k\Lambda^{k,n})Q_j^m + V_{ij}Q_j^{mn}$$
$$+ (V_{ik}^j Q_k^n + V_i^{jk}\Lambda^{k,n})\Lambda^{j,m} + V_i^j\Lambda^{j,mn} = 0 \qquad (2.2.5)$$

The third, fourth, etc. order equations can similarly be generated by successive perturbations of (2.2.3).

In applications of this procedure to specific problems, it is much more convenient to choose the η^k *a priori* from the basic variables, and in highly nonlinear cases the Q_i are often more suitable than the Λ^j. This is particularly so when the stability boundary is associated with *general* critical points. Nevertheless, the loading parameters yield an easier treatment, and the stability boundary associated with special critical points can be estimated conveniently by setting $\eta^k \equiv \Lambda^k$ in moderately nonlinear problems.

Allowing now M basic variables, appropriately chosen from the Q_i and Λ^j, to take on the role of the η^k, one observes that (2.2.4) represents a set of $(N \times M)$ linear equations in $(N \times M)$ first derivatives. If, for example, the Λ^j are equated to η^j, then, equation (2.2.4) reads

$$V_{ij}Q_j^k + V_i^k = 0; \qquad i, j = 1, 2, \ldots, N; \qquad (2.2.6)$$
$$k = 1, 2, \ldots, M$$

2.2.1 Estimation of the stability boundary

since $\Lambda^{j,k} = \delta^{jk}$ (Kronecker's delta), the Λ^j now being independent variables. In structural problems V_{ij} is positive definite at the origin and (2.2.6), therefore, yields a unique set of Q_j^k, indicating a unique equilibrium surface through the origin. This is true not only for the origin but for all noncritical equilibrium points, a fact which was observed earlier in another way.

Having evaluated the first derivatives, one substitutes them into the second order equation (2.2.5), which, then, represents a linear set of equations in the second derivatives Q_i^{mn} and $\Lambda^{j,mn}$ which can be solved for these derivatives. The aid of a digital computer will normally be needed for the solution of these linear sets of equations. Having determined the second derivatives, they are substituted in the third order equations which are, then, solved for the third derivatives and so on.

Having found the surface derivatives sequentially, the equilibrium surface can be constructed in series form to the desired degree of approximation as

$$Q_i = Q_i^m \eta^m + \frac{1}{2!} Q_i^{mn} \eta^m \eta^n + \cdots \tag{2.2.7}$$

$$\Lambda^j = \Lambda^{j,m} \eta^m + \frac{1}{2!} \Lambda^{j,mn} \eta^m \eta^n + \cdots \tag{2.2.8}$$

The fundamental surface (2.2.7 & 8) is normally stable initially. Any loss of stability, as one goes further from the origin on the surface, can be identified by the vanishing of the stability determinant

$$\det |V_{ij}(Q_k, \Lambda^l)| = 0 \tag{2.2.9}$$

The points lying on the equilibrium surface and satisfying (2.2.9) constitute the *critical zone* as defined in Section 2.1.2. The specific part of the critical zone which comprises the points lying at positions closest to the origin along the surface defines the stability boundary when projected into the load-space. The stability boundary associated with the *general* critical points can often be identified in the equilibrium analysis on the basis of (2.2.7 & 8) as the extrema on the surface, and although the following method is applicable to both cases, general or special alike, the latter for which the Λ^k can take the role of the η^k will be emphasized.

Consider now the variation of the stability determinant

$$\Delta(Q_k, \Lambda^l) \equiv \det |V_{ij}(Q_k, \Lambda^l)| \tag{2.2.10}$$

along the fundamental surface (2.2.7 & 8). The determinant along this surface is expressed as a function of the η^m,

$$A(\eta^m) \equiv \Delta[Q_i(\eta^m), \Lambda^j(\eta^m)] \tag{2.2.11}$$

2.2 General critical points

which can be differentiated and evaluated at the origin successively to obtain the ordered derivatives of A. Thus,

$$A^m = \Delta_i Q_i^m + \Delta^j \Lambda^{j,m},$$
$$A^{mn} = (\Delta_{ij} Q_j^n + \Delta_i^j \Lambda^{j,n}) Q_i^m + \Delta_i Q_i^{mn}$$
$$+ (\Delta_i^j Q_i^n + \Delta^{jk} \Lambda^{k,n}) \Lambda^{j,m} + \Delta^j \Lambda^{j,mn}, \qquad (2.2.12)$$
$$A^{mnr} = \cdots$$

It is seen that the surface derivatives Q_i^m, $\Lambda^{j,m}$, etc. determined in the equilibrium analysis can now be used readily to obtain A^m, A^{mn}, etc. The derivatives of the stability determinant, Δ_i, Δ^j, etc. are obtained by differentiating the determinant (2.2.10) and evaluating at the origin.

If the Λ^j are given the role of the η^j, then (2.2.12) reads

$$A^k = \Delta_i Q_i^k + \Delta^k$$
$$A^{kl} = (\Delta_{ij} Q_j^l + \Delta_i^l) Q_i^k + \Delta_i Q_i^{kl}$$
$$+ (\Delta_i^k Q_i^l + \Delta^{kl}) \qquad (2.2.13)$$

etc.

The stability determinant referred to the origin can now be written in the parametric form

$$A(\eta^m) = A_0 + A^m \eta^m + \frac{1}{2!} A^{mn} \eta^m \eta^n + \cdots \qquad (2.2.14)$$

Truncations of the right-hand side of (2.2.14) after the second, third, etc. terms yield the first, second, etc. order stability equations

$$A_0 + A^m \eta^m = 0$$
$$A_0 + A^m \eta^m + \frac{1}{2!} A^{mn} \eta^m \eta^n = 0 \qquad (2.2.15)$$

which are, then, solved with the equilibrium equations concurrently to obtain the first, second, etc. estimates for the stability boundary.

It is, finally, noted that in the event of Λ^j being chosen as the independent perturbation parameters, (2.2.15) yields the stability boundary directly as

$$A_0 + A^k \Lambda^k = 0$$
$$A_0 + A^k \Lambda^k + \frac{1}{2!} A^{kl} \Lambda^k \Lambda^l = 0 \qquad (2.2.16)$$

etc.

in which A^k, A^{kl}, etc. are given by (2.2.13).

In discussing the buckling behaviour of symmetric systems and in the analysis of the illustrative example arch, the role of the η^k will be shared by the loading parameters and the generalized coordinates.

2.2.2 Equilibrium surface in the vicinity of a general critical point

2.2.2 Equilibrium surface in the vicinity of a general critical point

Consider a structural system described by the total potential energy function $H = H(u_i, \lambda^j)$ as given by (2.1.7), and suppose the fundamental state F coincides with a *general* critical point at which

$$H_{11} = 0, \quad H_{ss} \neq 0 \quad \text{for all} \quad s \neq 1$$

and (2.2.17)

$$\overrightarrow{\text{grad}}_\lambda H_1(0, 0) \neq 0$$

Introduce the orthogonal transformation

$$\lambda^i = \beta^{ij}\varphi^j, \quad \beta^{ij}\beta^{ik} = \delta^{ik}, \quad (2.2.18)$$

to obtain a canonical representation [85] for the linear form corresponding to H_1^i, such that when (2.2.18) is substituted into the energy function $H(u_i, \lambda^j)$, the resulting transformed energy function

$$\Pi(u_i, \varphi^j) \equiv H(u_i, \beta^{ij}\varphi^j) \quad (2.2.19)$$

has the following properties:

$$\Pi_1^1(0, 0) \neq 0, \quad \Pi_1^m(0, 0) = 0 \quad \text{for all} \quad m \neq 1 \quad (2.2.20)$$

in the self-evident notation.

Clearly, this transformation is based on the assumption that some coefficient $H_1^i(0, 0)$ is not zero, and indeed, this condition is obviously satisfied by the very definition of the general critical point. It is also noted that since the transformation (2.2.18) is related to the λ^i only, the quadratic form of the new function $\Pi(u_i, \varphi^j)$ in the u_i is still diagonalized, i.e.,

$$\Pi_{ij}(0, 0) = 0 \quad \text{for} \quad i \neq j \quad (2.2.21)$$

and the general critical point is now defined simply by $\Pi_{11} = 0$ and $\Pi_{ss} \neq 0$ *for all* $s \neq 1$.

The analysis can now be based on the new potential energy function $\Pi(u_i, \varphi^j)$, the only necessary properties of which are given by the equations (2.2.20) and (2.2.21). It is not difficult to ascertain that the axes associated with the new parameters φ^m ($m = 2, 3, \ldots, M$) as well as u_1 are now tangential to the equilibrium surface, and these variables emerge, therefore, as ideal perturbation parameters. Thus, choosing u_1 and φ^m ($m = 2, 3, \ldots, M$) as M independent loading parameters, the solution of the equilibrium equations $\Pi_i = 0$ in the vicinity of F is expressed as

$$u_s = u_s(u_1, \varphi^m), \quad \varphi^1 = \varphi^1(u_1, \varphi^m) \quad (2.2.22)$$

2.2 General critical points

where $s \neq 1$, $m \neq 1$. Substituting these functions back into the equilibrium equations yields

$$\Pi_i[u_j(u_1, \varphi^m), \varphi^k(u_1, \varphi^m)] \equiv 0 \qquad (2.2.23)$$

in which $u_j(u_1, \varphi^m)$ and $\varphi^k(u_1, \varphi^m)$ are understood to reduce to u_1 and φ^m for $j = 1$ and $k = m \neq 1$, respectively.

The familiar perturbation pattern now follows. Thus, differentiating (2.2.23) once with respect to u_1 and once with respect to φ^m ($m = 2, 3, \ldots, M$), one gets

$$\Pi_{ij}u_{j,1} + \Pi_i^j \varphi_1^j = 0$$

and $\qquad (2.2.24)$

$$\Pi_{ij}u_j^m + \Pi_i^j \varphi^{j,m} = 0$$

respectively. Evaluating (2.2.24) at the critical point F results in [1]

$$\left.\begin{array}{r}\varphi_1^1 = 0 \\ \varphi^{1,m} = 0\end{array}\right\} \text{ for } i = 1 \qquad (2.2.25)$$

and

$$\left.\begin{array}{r}u_{s,1} = 0 \\ u_s^m = -\dfrac{\Pi_s^m}{\Pi_{ss}}\end{array}\right\} \text{ for } i = s \qquad (2.2.26)$$

Differentiating the first of the equations (2.2.24) with respect to u_1, one has

$$(\Pi_{ijk}u_{k,1} + \Pi_{ij}^k \varphi_1^k)u_{j,1} + \Pi_{ij}u_{j,11}$$
$$+ (\Pi_{ik}^j u_{k,1} + \Pi_i^{jk}\varphi_1^k)\varphi_1^j + \Pi_i^j \varphi_{11}^j = 0 \qquad (2.2.27)$$

which when evaluated at F yields

$$\varphi_{11}^1 = -\frac{\Pi_{111}}{\Pi_1^1} \quad \text{for} \quad i = 1$$

and $\qquad (2.2.28)$

$$u_{s,11} = -\frac{1}{\Pi_{ss}}\left(\Pi_{s11} - \frac{\Pi_s^1}{\Pi_1^1}\Pi_{111}\right) \quad \text{for} \quad i = s.$$

---

[1] It is noted that

$$\varphi^{1,1} = \frac{\partial \varphi^1}{\partial \varphi^1} = 1, \qquad \varphi^{1,11} = \frac{\partial^2 \varphi^1}{\partial (\varphi^1)^2} = 0, \text{ etc.}$$

$$u_{1,1} = \frac{\partial u_1}{\partial u_1} = 1, \qquad u_{1,11} = \frac{\partial^2 u_1}{\partial u_1^2} = 0, \text{ etc.}$$

2.2.2 Equilibrium surface in the vicinity of a general critical point

Differentiating again the first of equations (2.2.24) this time with respect to φ^m ($m = 2, 3, \ldots, M$), one has

$$(\Pi_{ijk} u_k^m + \Pi_{ij}^k \varphi^{k,m}) u_{j,1} + \Pi_{ij} u_{j,1}^m$$
$$+ (\Pi_{ik}^j u_k^m + \Pi_i^{jk} \varphi^{k,m}) \varphi_1^j + \Pi_i^j \varphi_1^{j,m} = 0 \quad (2.2.29)$$

which, upon evaluation at F, yields

$$\varphi_1^{1,m} = -\frac{1}{\Pi_1^1}\left(\Pi_{11}^m - \Pi_{11s}\frac{\Pi_s^m}{\Pi_{ss}}\right). \quad (2.2.30)$$

Now, differentiate the second of the equations (2.2.24) with respect to φ^n ($n = 2, 3, \ldots, M$) to get

$$(\Pi_{ijk} u_k^n + \Pi_{ij}^k \varphi^{k,n}) u_j^m + \Pi_{ij} u_j^{mn}$$
$$+ (\Pi_{ik}^j u_k^n + \Pi_i^{jk} \varphi^{k,n}) \varphi^{j,m} + \Pi_i^j \varphi^{j,mn} = 0 \quad (2.2.31)$$

which upon evaluation and, of course, using previously obtained derivatives (2.2.25) and (2.2.26), yields

$$\varphi^{1,mn} = -\frac{1}{\Pi_1^1}\left(\Pi_1^{mn} + \Pi_{1sr}\frac{\Pi_s^m \Pi_r^n}{\Pi_{ss}\Pi_{rr}} - \frac{\Pi_{1s}^n \Pi_s^m + \Pi_{1s}^m \Pi_s^n}{\Pi_{ss}}\right) \quad (2.2.32)$$

where $s \neq 1$, $r \neq 1$, $m \neq 1$ and $n \neq 1$.

Asymptotic relationships for $\varphi^1 = \varphi^1(u_1, \varphi^m)$ and $u_s = u_s(u_1, \varphi^m)$, defining the equilibrium surface in the vicinity of F, can now be constructed. Thus, using Taylor's expansion and the derivatives (2.2.25), (2.2.26), (2.2.28), (2.2.30) and (2.2.32), one has

$$\Pi_1^1 \varphi^1 + \tfrac{1}{2}\Pi_{111} u_1^2 + \left(\Pi_{11}^m - \Pi_{11s}\frac{\Pi_s^m}{\Pi_{ss}}\right) u_1 \varphi^m$$
$$+ \frac{1}{2}\left(\Pi_1^{mn} + \Pi_{1sr}\frac{\Pi_s^m \Pi_r^r}{\Pi_{ss}\Pi_{rr}} - 2\frac{\Pi_{1s}^m \Pi_s^n}{\Pi_{ss}}\right)\varphi^m \varphi^n = 0 \quad (2.2.33)$$

and

$$\Pi_{ss} u_s + \Pi_s^m \varphi^m + \frac{1}{2}\left(\Pi_{s11} - \frac{\Pi_s^1}{\Pi_1^1}\Pi_{111}\right) u_1^2 = 0 \quad (2.2.34)$$

Substituting for u_1 from (2.2.33), equation (2.2.34) can also be written as

$$\Pi_{ss} u_s + \Pi_s^m \varphi^m + \left(\Pi_s^1 - \frac{\Pi_{s11}\Pi_1^1}{\Pi_{111}}\right)\varphi^1 = 0 \quad (2.2.35)$$

The equations (2.2.33) and (2.2.35), when considered together, define the M dimensional equilibrium surface in the vicinity of the general critical point F, and when considered individually they represent the

2.2 General critical points

projections of this surface into the $u_1-\varphi^i$ and $u_s-\varphi^m$ subspaces, respectively.

Consider now, a general ray given by $\varphi^i = l^i\xi$ where at least $l^1 \neq 0$; the equilibrium equations (2.2.33) and (2.2.35), then, yield

$$\Pi_1^1 l^1 \xi + \tfrac{1}{2}\Pi_{111} u_1^2 = 0 \qquad (2.2.36)$$

and

$$\Pi_{ss} u_s + \left[\Pi_s^m l^m + \left(\Pi_s^1 - \frac{\Pi_{s11}\Pi_1^1}{\Pi_{111}}\right) l^1\right]\xi = 0 \qquad (2.2.37)$$

which indicate a limit point on a plot of u_1 against ξ.

On the other hand, suppose one takes a special ray such that $l^1 = 0$ and $l^m \neq 0$ ($m \neq 1$); the equilibrium equations, then, yield

$$u_1 = \frac{1}{\Pi_{111}}[-a \pm (a^2 - \Pi_{111} b)^{\frac{1}{2}}]\xi \qquad (2.2.38)$$

and

$$u_s = -\frac{\Pi_s^m l^m}{\Pi_{ss}}\xi \qquad (2.2.39)$$

in which a and b are constants:

$$\begin{aligned}a &= \left(\Pi_{11}^m - \Pi_{s11}\frac{\Pi_s^m}{\Pi_{ss}}\right) l^m \equiv c^m l^m \\ b &= \left(\Pi_1^{mn} + \Pi_{1sr}\frac{\Pi_s^m \Pi_r^n}{\Pi_{ss}\Pi_{rr}} - 2\frac{\Pi_{1s}^m \Pi_s^n}{\Pi_{ss}}\right) l^m l^n \equiv d^{mn} l^m l^n\end{aligned} \qquad (2.2.40)$$

Obviously, as discussed in Section 2.1.3, (2.2.38) indicates bifurcation provided $a^2 - \Pi_{111} b > 0$.

In order to visualize the equilibrium surface, consider the special case of $M = 2$, reducing the parameters to two, namely, φ^1 and φ^2; then, (2.2.33) becomes the equation of a

synclastic	< 0	(2.2.41a)
anticlastic surface if $a^2 - \Pi_{111} b > 0$		(2.2.41b)
parabolic	$= 0$	(2.2.41c)

These surfaces are illustrated in Figures 2.7, 2.8 and 2.9, respectively. In this terminology, the critical point F can be called *elliptic*, *hyperbolic* or *parabolic* according to whether (2.2.41a), (2.2.41b) or (2.2.41c) holds respectively.

It is clear that in the event of elliptic and parabolic points, bifurcation is ruled out while a hyperbolic point can be regarded as a limit point on a plot of u_1 versus φ^1 and a bifurcation point on a $u_1 - \varphi^m$ plot.

2.2.2 Equilibrium surface in the vicinity of a general critical point

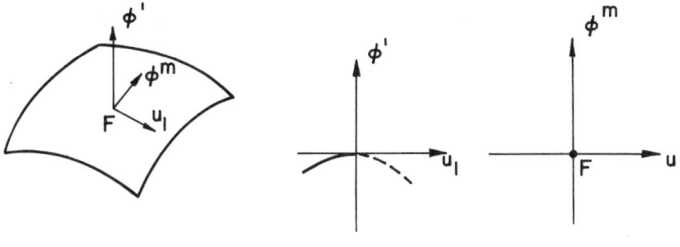

Figure 2.7

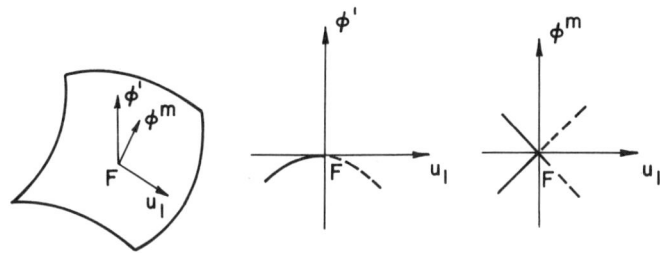

Figure 2.8

Extending the terminology used for two-dimensional surfaces to multi-dimensional surfaces, one can call the surface (2.2.33) *synclastic* if (2.2.41a) holds for *all* possible rays given by $\varphi^i = l^i \xi$ where $l^1 = 0$, and some $l^m \neq 0$ ($m \neq 1$) in which case the matrix

$$[c^m c^n - \Pi_{111} d^{mn}] \tag{2.2.42}$$

is negative definite. The surface (2.2.33) will be called *anticlastic* if (2.2.41b) holds for *all* rays defined above in which case the matrix (2.2.42) is positive definite. Similarly, the surface (2.2.33) will be called *parabolic* if (2.2.41c) holds for all rays defined above in which case the matrix

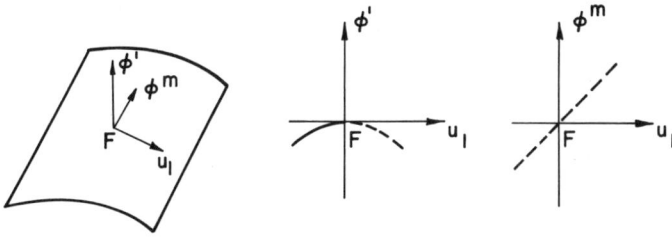

Figure 2.9

2.2 General critical points

(2.2.42) becomes a null matrix. Clearly, if the surface is not synclastic, it does not necessarily follow that it is either anticlastic or parabolic. In fact, when the matrix (2.2.42) is indefinite, the surface takes quite an arbitrary shape. One of the reasons for introducing the above terminology is to simplify referring to these particular shapes which will often arise. Finally, it must be noted that the parabolic case will often exhibit some sort of curvature in higher order approximations.

So far, it has been assumed that the energy coefficients (excluding the necessary properties of Π) are nonzero. It was shown earlier, however, that the cubic coefficient H_{111} plays an important role in describing the buckling behaviour of the system, and the general critical point is called *singular* if this coefficient vanishes. Suppose now that $\Pi_{111} = 0$; then, the derivative φ^1_{11} vanishes and one proceeds to differentiate the equation (2.2.27) with respect to u_1 once more to get

$$\varphi^1_{111} = -\frac{\bar{\Pi}_{1111}}{\Pi^1_1} \equiv -\frac{1}{\Pi^1_1}\left[\Pi_{1111} - 3\sum_{s=2}^{N}\frac{(\Pi_{s11})^2}{\Pi_{ss}}\right] \tag{2.2.43}$$

after evaluation at F for $i = 1$.

The first order equation of the equilibrium surface in the $u_1 - \varphi^i$ space follows as

$$\Pi^i_1\varphi^1 + \frac{1}{3!}\bar{\Pi}_{1111}u^3_1 + c^m u_1\varphi^m + \tfrac{1}{2}d^{mn}\varphi^m\varphi^n = 0 \tag{2.2.44}$$

The discussion related to equation (2.1.29) obviously applies to the equation (2.2.44) as well, and will not be repeated. The projection of the equilibrium surface into $u_s - u_1 - \varphi^m$ space is obtained by setting $\Pi_{111} = 0$ in (2.2.34) as

$$\Pi_{ss}u_s + \Pi^m_s\varphi^m + \tfrac{1}{2}\Pi_{s11}u^2_1 = 0$$

Note that in this case u_s can only be expressed in terms of u_1 and φ^m.

2.2.3 Critical zone and the stability boundary

In the preceding section the shape of the equilibrium surface in the vicinity of a general critical point, F, has been investigated. In this section, the surface is explored further to locate other critical points which might exist on the surface in the vicinity of F.

Critical points should satisfy the determinantal equation

$$\Delta(u_k, \varphi^i) \equiv \det|\Pi_{ij}(u_k, \varphi^i)| \equiv \det|H_{ij}(u_k, \lambda^i)| = 0 \tag{2.2.45}$$

Differentiating the determinant $\Delta(u_k, \varphi^i) \equiv |\Pi_{ij}(u_k, \varphi^i)|$ once with respect to u_i ($i = 1, 2, \ldots, N$) and once with respect to φ^j ($j = 1, 2, \ldots, M$),

2.2.3 Critical zone and the stability boundary

and evaluating at F one gets

$$\Delta_i = \Pi_{11i} \prod_{s=2}^{N} \Pi_{ss}$$
$$\Delta^j = \Pi_{11}^j \prod_{s=2}^{N} \Pi_{ss} \tag{2.2.46}$$

which will be used later in the analysis. It is assumed for now that F is not singular; discussion of singular critical points is deferred to Section 2.2.6.

Considering now the equation (2.2.45) together with the equilibrium equations, one observes that the critical states of equilibrium (i.e., the *critical zone*) can be expressed in the form of $N+1$ functions of $M-1$ independent parameters which are chosen here as the φ^m ($m = 2, \ldots, M$). Thus, the critical zone is assumed to be in the form

$$\overset{*}{u}_i = \overset{*}{u}_i(\overset{*}{\varphi}{}^m), \qquad \overset{*}{\varphi}{}^1 = \overset{*}{\varphi}{}^1(\overset{*}{\varphi}{}^m) \tag{2.2.47}$$

where a star (*) is used to denote the critical variables. If these functions are substituted back into the equilibrium equations, $\Pi_i = 0$, and the criticality condition $\Delta = 0$, the identities

$$\Pi_i[\overset{*}{u}_i(\overset{*}{\varphi}_m), \overset{*}{\varphi}{}^1(\overset{*}{\varphi}{}^m), \overset{*}{\varphi}{}^m] \equiv 0$$
$$\Delta[\overset{*}{u}_i(\overset{*}{\varphi}{}^m), \overset{*}{\varphi}{}^1(\overset{*}{\varphi}{}^m), \overset{*}{\varphi}{}^m] \equiv 0 \tag{2.2.48}$$

are obtained.

Differentiating (2.2.48) with respect to $\overset{*}{\varphi}{}^m$ ($m = 2, 3, \ldots, M$) yields

$$\Pi_{ij}\overset{*}{u}{}^m_j + \Pi^1_i\overset{*}{\varphi}{}^{1,m} + \Pi^m_i = 0$$
$$\Delta_i\overset{*}{u}{}^m_i + \Delta^1\overset{*}{\varphi}{}^{1,m} + \Delta^m = 0 \tag{2.2.49}$$

which, upon evaluation at F, result in

$$\overset{*}{\varphi}{}^{1,m} = 0$$
$$\overset{*}{u}{}^m_s = -\frac{\Pi^m_s}{\Pi_{ss}} \tag{2.2.50}$$
$$\overset{*}{u}{}^m_1 = -\frac{1}{\Pi_{111}}\left(\Pi^m_{11} - \Pi_{s11}\frac{\Pi^m_s}{\Pi_{ss}}\right)$$

Differentiating the first of equations (2.2.49) with respect to $\overset{*}{\varphi}{}^n$ ($n = 2, \ldots, M$) one obtains

$$(\Pi_{ijk}\overset{*}{u}{}^n_k + \Pi^1_{ij}\overset{*}{\varphi}{}^{1,n} + \Pi^n_{ij})\overset{*}{u}{}^m_j + \Pi_{ij}\overset{*}{u}{}^{mn}_j$$
$$+ (\Pi^1_{ij}\overset{*}{u}{}^n_j + \Pi^{11}_i\overset{*}{\varphi}{}^{1,n} + \Pi^{1n}_i)\overset{*}{\varphi}{}^{1,m} + \Pi^1_i\overset{*}{\varphi}{}^{1,mn}$$
$$+ \Pi^m_{ij}\overset{*}{u}{}^n_j + \Pi^{m1}_i\overset{*}{\varphi}{}^{1,n} + \Pi^{mn}_i = 0 \tag{2.2.51}$$

2.2 General critical points

Evaluation at F yields

$$\overset{*}{\varphi}{}^{1,mn} = \frac{1}{\Pi_1^1 \Pi_{111}} (c^m c^n - \Pi_{111} d^{mn}) \qquad (2.2.52)$$

in which c^m and d^{mn} are given by (2.2.40).

The asymptotic equations of the *critical zone* can now be constructed readily as

$$\overset{*}{\varphi}{}^1 = \frac{1}{2\Pi_1^1 \Pi_{111}} (c^m c^n - \Pi_{111} d^{mn}) \overset{*}{\varphi}{}^m \overset{*}{\varphi}{}^n \qquad (2.2.53)$$

$$\overset{*}{u}_1 = -\frac{1}{\Pi_{111}} \left(\Pi_{11}^m - \Pi_{s11} \frac{\Pi_s^m}{\Pi_{ss}} \right) \overset{*}{\varphi}{}^m \equiv -\frac{1}{\Pi_{111}} c^m \overset{*}{\varphi}{}^m \qquad (2.2.54)$$

$$\overset{*}{u}_s = -\frac{\Pi_s^m}{\Pi_{ss}} \overset{*}{\varphi}{}^m \qquad (2.2.55)$$

· The equation (2.2.53) represents the *stability boundary* (in general the *critical surfaces*) if the critical point F is *primary* and consequently all $H_{ss} > 0$. Equations (2.2.54) and (2.2.55) represent the projections of the critical zone into $u_1 - \varphi^m$ and $u_s - \varphi^m$ subspaces respectively.

Now suppose the critical state F is an *elliptic* point satisfying (2.2.41a) for all rays given by $\varphi^m = l^m \xi$, then, the matrix $[c^m c^n - \Pi_{111} d^{mn}]$ is negative definite, and consequently the stability boundary (2.2.53) is a *synclastic* (*strictly convex*) surface *concave* towards the region of existence which is identified by the curvature $\varphi_{11}^1|_F = -\Pi_{111}/\Pi_1^1$ obtained from the equilibrium equation (2.2.33) on a plot of u_1 against φ^1. In the majority of practical problems, the stability boundary becomes concave towards the origin of the loading space (Figure 2.10).

On the other hand, if the fundamental state F is a *hyperbolic* point, it follows from the definition of such a point and (2.2.53) that the stability boundary is again synclastic but this time *convex* towards the region of existence while the equilibrium surface as a whole is anticlastic (Figure 2.11). Finally, if F is a parabolic point, the stability boundary becomes a plane (Figure 2.12).

It must be noted that, in general, the stability boundary might not be synclastic, and this happens when the matrix $[c^m c^n - \Pi_{111} d^{mn}]$ is *indefinite*; this possibility, however, is ruled out in case $M = 2$, involving only two parameters, φ^1 and φ^2.

Clearly, if the matrix $[c^m c^n - \Pi_{111} d^{mn}]$ is negative definite, that is,

$$a^2 - \Pi_{111} b < 0 \quad \text{for all} \quad \varphi^m = l^m \xi \qquad (2.2.56)$$

the stability boundary cannot have convexity towards the region of existence, and since the condition (2.2.56) also rules out the possibility of bifurcation, the following theorem is established:

2.2.3 Critical zone and the stability boundary

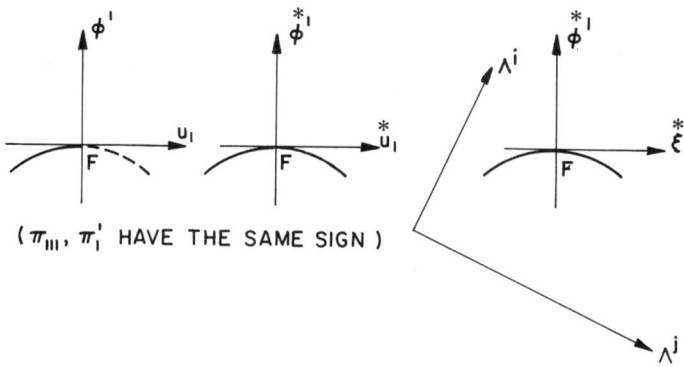

Figure 2.10

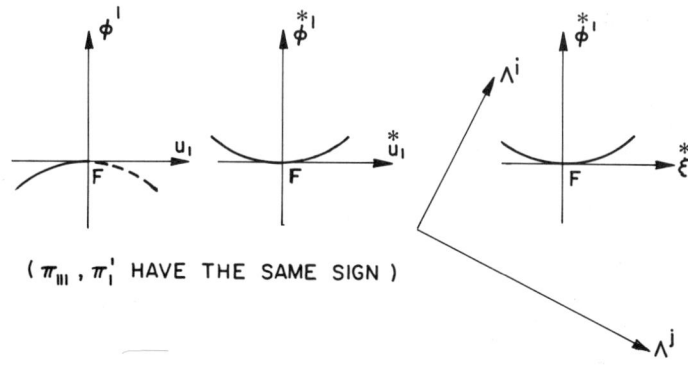

Figure 2.11

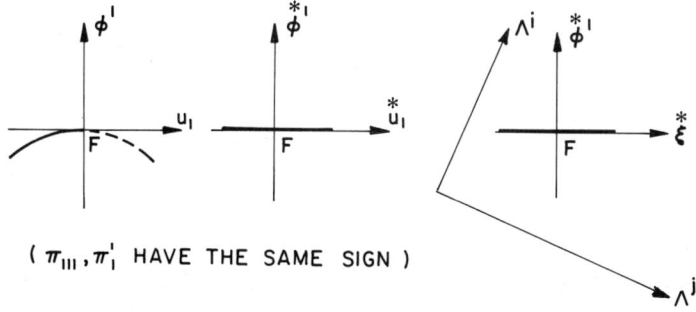

Figure 2.12

2.2 General critical points

THEOREM 2.1. *If the general critical point, F, is elliptic at which the possibility of bifurcation is ruled out, then, the associated stability boundary cannot have convexity towards the region of existence.*

COROLLARY 1. *A necessary condition that the stability boundary is convex towards the region of existence is that bifurcation of solution at F is possible.*

COROLLARY 2. *In the particular case of two parameters (φ^1 and φ^2) only, the necessary condition of Corollary 1 becomes also sufficient; i.e., if bifurcation at F can occur, then, the stability boundary is convex with respect to the region of existence.*

2.2.4 The critical surface as the existence boundary

It is noted that the curvature $\varphi^1_{11}|_F = -\dfrac{\Pi_{111}}{\Pi^1_1}$ is being used as a criterion to determine the regions of existence and inexistence; this point, however, needs further clarification.

Evidently, if $\varphi^m = 0$ ($m \neq 1$), real equilibrium states in the vicinity of F can only exist for either $\varphi^1 > 0$ or $\varphi^1 < 0$ depending on the sign of the coefficients Π_{111} and Π^1_1. It can be shown that the critical surfaces (and in particular the stability boundary) constitute an existence boundary in the sense that real equilibrium states can only exist for the points (of the load-space) lying on one side of the critical surface, there being no *neighbouring* equilibrium states corresponding to points lying on the other side of the critical surface.

To this end, consider an arbitrarily chosen point A on the critical surface defined by a set of critical parameters $\overset{*}{\varphi}{}^i_A$ ($i = 1, 2, \ldots, M$), and examine the equilibrium states corresponding to

$$\varphi^m = \overset{*}{\varphi}{}^m_A, \quad m \neq 1$$
$$\varphi^1 = \overset{*}{\varphi}{}^1_A + \tau \tag{2.2.57}$$

where τ is a small increment in $\overset{*}{\varphi}{}^1_A$. For a certain value of τ, (2.2.57) defines a point in the original load-space (Λ^i-space).

Substituting for φ^1 and φ^m in the equilibrium equation (2.2.33) yields

$$\Pi^1_1 \overset{*}{\varphi}{}^1_A + \Pi^1_1 \tau + \tfrac{1}{2}\Pi_{111} u^2_1 + c^m u_1 \overset{*}{\varphi}{}^m_A + \tfrac{1}{2} d^{mn} \overset{*}{\varphi}{}^m_A \overset{*}{\varphi}{}^n_A = 0$$

Using equations (2.2.53) and (2.2.54), which must be satisfied by the critical equilibrium states in the vicinity of F, one has

$$u_1 = \overset{*}{u}{}^A_1 \pm \frac{1}{\Pi_{111}} (-2\Pi_{111} \Pi^1_1 \tau)^{\frac{1}{2}} \tag{2.2.58}$$

2.2.4 The critical surface as the existence boundary

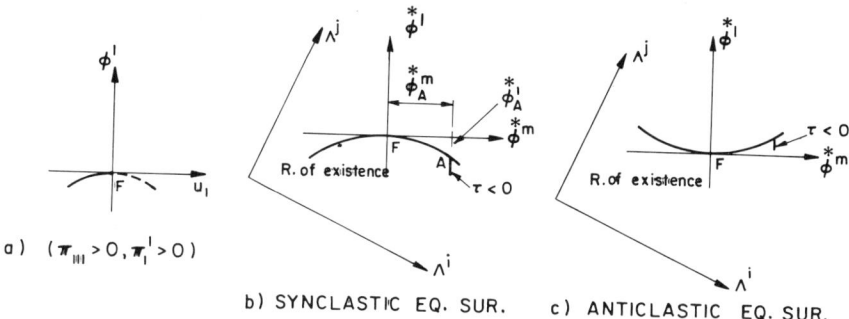

Figure 2.13

Suppose now that Π_{111} and $\Pi_1^!$ have the same sign, then, real equilibrium states can only exist for $\tau < 0$. As the point A slides on the critical surface, $\tau < 0$ defines the *region of existence* and $\tau > 0$ (the other side of the critical surface) defines the *region of inexistence*. On the other hand, if Π_{111} and $\Pi_1^!$ have the opposite signs, real solutions can only exist for $\tau > 0$ defining the corresponding region of existence. Obviously $\tau = 0$ gives the point A.

Thus it can be concluded that the critical surface is, indeed, an existence boundary and regions of existence and inexistence can be located by examining the sign of the curvature $\varphi_{11}^1|_F$. This is a general result which is valid regardless of the shape of the equilibrium surface, and the following theorem can be stated:

THEOREM 2.2. *The critical surface constitutes an existence boundary so that neighbouring equilibrium states can exist only for the points (of the load-space) lying on one side of the critical surface.*

Figures 2.13 and 2.14 illustrate this phenomenon with regard to synclastic and anticlastic surfaces.

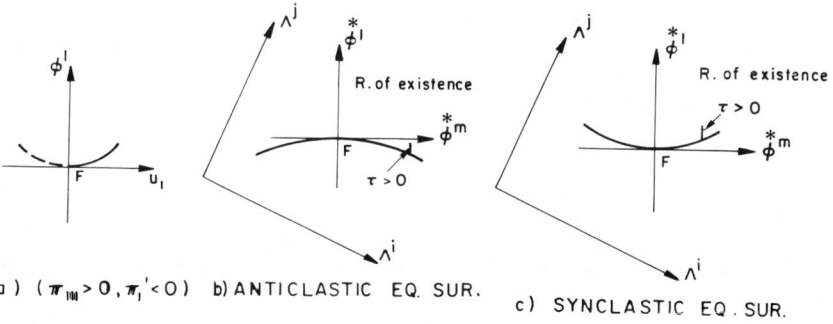

Figure 2.14

137

2.2 General critical points

2.2.5 One-degree-of-freedom systems

So far, the potential energy function V and consequently the functions H and Π have been treated as general functions in the sense that they were not assumed to be linear in the loading parameters. It has been remarked earlier, however, that the energy function is often linear in the parameters λ^j and in particular when these parameters represent the external loads. Then, on the basis of the transformation (2.2.18), it is concluded that the second and higher order derivatives of the function Π with respect to the φ^i vanish, i.e.,

$$\Pi_i^{jk} = \Pi_i^{jkl} = \cdots = \Pi_{ij}^{kl} = \Pi_{ij}^{klm} = \cdots = 0 \qquad (2.2.59)$$

It follows from (2.2.53) and (2.2.33) that the stability boundary for a one-degree-of-freedom system takes the form

$$\overset{*}{\varphi}{}^1 = \frac{\Pi_{11}^m \Pi_{11}^n}{2\Pi_1^1 \Pi_{111}} \overset{*}{\varphi}{}^m \overset{*}{\varphi}{}^n \equiv \frac{(\Pi_{11}^m \overset{*}{\varphi}{}^m)^2}{2\Pi_1^1 \Pi_{111}} \qquad (2.2.60)$$

and the equilibrium surface is given by

$$\Pi_i^1 \varphi^1 + \tfrac{1}{2}\Pi_{111} u_1^2 + \Pi_{11}^m u_1 \varphi^m = 0 \qquad (2.2.61)$$

Assuming further that there are only two loading parameters, namely φ^1 and φ^2, one observes that the curvature of the stability boundary

$$\overset{*}{\varphi}{}^{1,22} = \frac{(\Pi_{11}^2)^2}{\Pi_1^1 \Pi_{111}}$$

and the curvature

$$\varphi_{11}^1 = -\frac{\Pi_{111}}{\Pi_1^1}$$

will always have opposite signs. It is, then, concluded that the stability boundary is convex towards the region of existence.

Returning now to the general case of M parameters, one observes that (2.2.60) can assume zero values for suitably chosen $\overset{*}{\varphi}{}^m$. This indicates that the stability boundary will be *flat* in certain directions to the degree of approximation maintained in the equation (2.2.60), and the question immediately arises as to the possibility of certain curvatures having opposite signs when the higher order derivatives are determined. It can, nevertheless, be proved that the stability boundary of a one-degree-of-freedom system cannot have convexity towards the region of inexistence and this possibility is ruled out. In order to establish such a stronger theorem, a slightly modified formulation is required, and will now be pursued due to its significance.

2.2.5 One-degree-of-freedom systems

Thus, starting with the energy function $H = H(u_1, \lambda^i)$ referred to the fundamental state F which is again chosen as a general critical point on the equilibrium surface at which $H_{11} = 0$ and $\overrightarrow{\text{grad}}_\lambda H_1(0, 0) \neq 0$, introduce the transformation

$$\lambda^i = \beta^{ij} \varphi^j, \quad \beta^{ij} \beta^{jk} = \delta^{ik}$$

such that when this is substituted into the energy function $H(u_1, \lambda^i)$ the resulting transformed energy function

$$T(u_1, \varphi^i) \equiv H(u_1, \beta^{ij} \varphi^j) \qquad (2.2.62)$$

has the following properties:

$$T_1^1(0,0) \neq 0, \quad T_1^m(0,0) = 0 \quad \text{for} \quad m = 2, 3, \ldots, M$$

and $\qquad (2.2.63)$

$$T_{11}^2(0,0) \neq 0, \quad T_{11}^r(0,0) = 0 \quad \text{for} \quad r = 3, 4, \ldots, M$$

Clearly, this transformation is based on the assumption that the two linear forms associated with $H_1^i(0, 0)$ and $H_{11}^i(0, 0)$ are linearly independent. The trivial case in which these forms become linearly dependent will be discussed later. Finally, it is noted that the property (2.2.59) is carried over to the function T which will be used in the following analysis.

Following the same approach as before the derivatives of the equilibrium surface can be obtained by successive perturbations of the identity

$$T_1[\varphi^1(u_1, \varphi^m), u_1, \varphi^m] \equiv 0 \qquad (2.2.64)$$

which results in

$$\varphi_1^1 = 0$$
$$\varphi^{1,m} = 0 \qquad (2.2.65)$$

$$\varphi_{11}^1 = -\frac{T_{111}}{T_1^1}$$
$$\varphi^{1,mn} = 0 \qquad (2.2.66)$$

and

$$\varphi_1^{1,2} = -\frac{T_{11}^2}{T_1^1} \qquad (2.2.67)$$
$$\varphi_1^{1,r} = 0 \quad \text{for} \quad r = 3, 4, \ldots, M$$

successively. Higher order perturbations show that all the derivatives of φ^1 with respect to φ^r ($r \neq 1, 2$) and u_1 vanish at F. It follows, therefore, that

2.2 General critical points

the equilibrium surface in the vicinity of F has the form

$$T_1^1\varphi^1 + T_{11}^2 u_1\varphi^2 + \tfrac{1}{2}T_{111}u_1^2 = 0 \qquad (2.2.68)$$

Similarly, the critical zone can be obtained by successive perturbations of

$$T_1[\overset{*}{\varphi}{}^1(\overset{*}{\varphi}{}^m), \overset{*}{u}_1(\overset{*}{\varphi}{}^m), \overset{*}{\varphi}{}^m] \equiv 0$$
$$T_{11}[\overset{*}{\varphi}{}^1(\overset{*}{\varphi}{}^m), \overset{*}{u}_1(\overset{*}{\varphi}{}^m), \overset{*}{\varphi}{}^m] \equiv 0 \qquad (2.2.69)$$

which yield the derivatives

$$\overset{*}{\varphi}{}^{1,m} = 0$$

$$\overset{*}{u}_1^2 = -\frac{T_{11}^2}{T_{111}} \qquad (2.2.70)$$

$$\overset{*}{u}_1^r = 0 \quad \text{for} \quad r = 3, 4, \ldots, M$$

and

$$\overset{*}{\varphi}{}^{1,22} = \frac{(T_{11}^2)^2}{T_1^1 T_{111}} \qquad (2.2.71)$$

$$\overset{*}{\varphi}{}^{1,2r} = 0$$

$$\overset{*}{\varphi}{}^{1,rs} = 0 \quad \text{for} \quad r, s = 3, 4, \ldots, M$$

Again, higher order perturbations do not produce finite derivatives of $\overset{*}{\varphi}{}^1$ with respect to $\overset{*}{\varphi}{}^r$ ($r \neq 1, 2$), and the same applies to the derivatives of $\overset{*}{u}_1$ with respect to $\overset{*}{\varphi}{}^r$ ($r \neq 1, 2$). Hence, the critical zone around F takes the simple form

$$T_{111}\overset{*}{u}_1 + T_{11}^2 \overset{*}{\varphi}{}^2 = 0$$

$$T_1^1\overset{*}{\varphi}{}^1 - \frac{(T_{11}^2)^2}{2T_{111}}(\overset{*}{\varphi}{}^2)^2 = 0, \qquad (2.2.72)$$

the latter equation defining the stability boundary. Comparing equations (2.2.72) with earlier results reveals that the transformation (2.2.63) has, in effect, reduced the system to one with two *significant* parameters, namely φ^1 and φ^2, the φ^r ($r \neq 1, 2$) being eliminated.

The curvatures $\varphi_{11}^1 = -T_{111}/T_1^1$ and $\overset{*}{\varphi}{}^{1,22} = (T_{11}^2)^2/T_1^1 T_{111}$ now show clearly that their signs will always be opposite, and the result obtained for two loads earlier can be generalized to M loads in the form of

LEMMA 1. *The stability boundary is convex with regard to the region of existence.*

The trivial case in which the linear forms associated with $H_1^1(0, 0)$ and $H_{11}^i(0, 0)$ are linearly dependent has yet to be examined. Obviously, the transformation (2.2.62) resulting in the properties (2.2.63) cannot be achieved. Instead, however, a similar transformation can be introduced such that

$$T_1^i(0,0) \neq 0, \quad T_1^m(0,0) = 0 \quad \text{for} \quad m = 2, 3, \ldots, M$$
$$T_{11}^1(0,0) \neq 0, \quad T_1^m(0,0) = 0 \quad \text{for} \quad m = 2, 3, \ldots, M \quad (2.2.73)$$

Following the same perturbation procedure as before, it is not difficult to construct the asymptotic equations of the equilibrium surface and the stability boundary as

$$T_1^1 \varphi^1 + \tfrac{1}{2} T_{111}(u_1)^2 = 0 \quad (2.2.74)$$

and

$$\overset{*}{\varphi}{}^1 = 0 \quad (2.2.75)$$

respectively. The obvious conclusion is:

LEMMA 2. *The stability boundary is a plane.*

Lemma 2 leads to an immediate

COROLLARY. *A sufficient condition that the stability boundary takes the form of a flat surface (plane) in the vicinity of a general critical point F is that the two linear forms associated with $H_1^i(0,0)$ and $H_{11}^i(0,0)$ are linearly dependent.*

Combining *Lemma 1* and *Lemma 2*, one can formulate the

THEOREM 2.3. *The stability boundary of a one-degree-of-freedom system cannot have convexity with regard to the region of inexistence (instability).*

This theorem is stronger than Theorem 2.1, in that it does not stipulate additional conditions, i.e., for one-degree-of-freedom systems, the general critical point cannot be *elliptic*.

2.2.6 Singular critical points

Through Sections 2.2.3 to 2.2.5 it was assumed that the critical point F was not singular in the sense that the cubic coefficient $\Pi_{111} \neq 0$. This coefficient appears in several surface derivatives, and if it vanishes the equations of the critical zone will have to be rederived. However, the formulation of Section 2.2.3 does not allow for such a derivation via the

2.2 General critical points

perturbation parameters $\overset{*}{\phi}{}^m$ ($m \neq 1$), since some of the derivatives, $\overset{*}{u}{}_1^m$ in (2.2.50) for example, tend to infinity as Π_{111} approaches zero. It seems that $\overset{*}{u}_1$ must be one of the perturbation parameters, and the analysis can be carried out by choosing u_1 and φ^r ($r = 3, 4, \ldots, M_1$) as independent, thus leading to the functions $\overset{*}{\phi}{}^1(\overset{*}{u}_1, \overset{*}{\phi}{}^r)$, $\overset{*}{\phi}{}^2(\overset{*}{u}_1, \overset{*}{\phi}{}^r)$, $\overset{*}{u}_s(\overset{*}{u}_1, \overset{*}{\phi}{}^r)$ to be determined.

Such an analysis, however, will not be pursued here; instead, the critical zone will be obtained by considering the extremum points on the surface (2.2.44) with respect to u_1. Thus, differentiating (2.2.44) with respect to u_1 and equating to zero yields the critical relationship

$$\tfrac{1}{2}\bar{\Pi}_{1111}\overset{*}{u}{}_1^2 + c^m \overset{*}{\phi}{}^m = 0 \qquad (2.2.76)$$

Substituting for u_1 into (2.2.44) and neglecting $d^{mn}\varphi^m\varphi^n$ compared to $(c^m\varphi^m)^{\frac{3}{2}}$, one obtains the stability boundary

$$c^m \overset{*}{\phi}{}^m = -\tfrac{1}{2}\bar{\Pi}_{1111}^{\frac{1}{3}}(3\Pi_1^1\overset{*}{\phi}{}^1)^{\frac{2}{3}} \qquad (2.2.77)$$

representing a generalization of the two-thirds power law associated with imperfect systems which was discussed in Part 1, [see Eq. (1.2.20)]. For each ray $\overset{*}{\phi}{}^m = l^m \overset{*}{\xi}$, then, equation (2.2.77) yields a sharp cusp on a plot of $\overset{*}{\xi}$ versus $\overset{*}{\phi}{}^1$, and if the bifurcation on the $\xi - u_1$ plot is unstable symmetric, the stability boundary in the vicinity of F takes the form shown in Figure 2.15.

It was seen that the point F on the equilibrium surface (2.2.44) appears as a symmetric point of bifurcation if a loading ray in the φ^m subspace is followed, and other singular critical points on the stability boundary (2.2.77) can, therefore, be identified by setting $\overset{*}{\phi}{}^1 = 0$ in (2.2.77), leading to

THEOREM 2.4. *Singular critical points in the vicinity of F form a plane.*

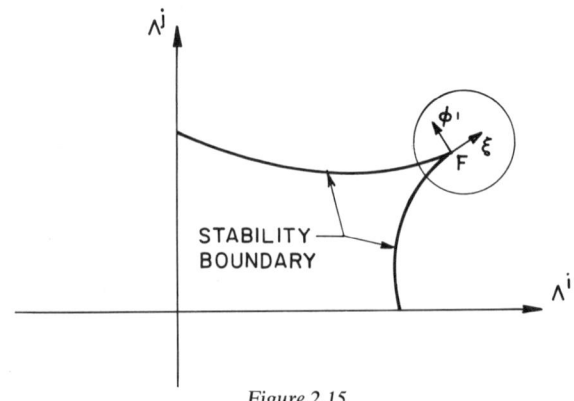

Figure 2.15

2.2.6 Singular critical points

It is seen that a singular critical point results in an abrupt change in the shape of the stability boundary, and for upper and/or lower bound estimates of the stability boundary, the location of such points must be known.

Consider first a one-degree-of-freedom system described initially by the potential energy function $V(Q_1, \Lambda^i)$ and suppose the location of singular critical points (if any) are sought. Assuming further that the total potential energy, referred to the origin is in the form

$$V(Q_1, \Lambda^i) = \frac{1}{2!} V_{11} Q_1^2 + \frac{1}{3!} V_{111} Q_1^3 + \cdots$$
$$+ V_1^i Q_1 \Lambda^i + \frac{1}{2!} V_{11}^i Q_1^2 \Lambda^i + \cdots, \qquad (2.2.78)$$

one attempts to find the rays $\Lambda^i = l^i \xi$ which lead to singular critical points, resulting in a symmetric point of bifurcation on a plot of ξ versus Q_1. Such a ray represents an appropriate combination of loads, which in most cases, keeps the structure undeflected until bifurcation occurs, i.e., external loads are combined such that, as the loading parameter ξ is increased gradually the structure maintains its unloaded configuration without undergoing deflections until a singular point is reached. Obviously, this can only happen if

$$V_1^i l^i = 0 \qquad (2.2.79)$$

reducing the system to a linear eigenvalue problem. The critical points will, then, be given by

$$V_{11} + V_{11}^i l^i \xi = 0 \qquad (2.2.80)$$

Equations (2.2.79) and (2.2.80) can also be written as

$$V_1^i \Lambda^i = 0$$

and $\qquad (2.2.81)$

$$V_{11} + V_{11}^i \Lambda^i = 0$$

which define a point if $M = 2$, a straight line if $M = 3$ and two or multi-dimensional planes if $M = 4$ or $M > 4$ respectively.

It is noted, however, that the condition $\partial^3 V/\partial Q_1^3|_{Q_1=0}$ has not been imposed yet, and the critical points given by (2.2.81) are not necessarily singular unless this condition is also satisfied. If this cubic coefficient does not vanish, the critical points obtained from (2.2.81) are nonsingular general points. The ray described by ξ is, then, in the direction of φ^m ($m \neq 1$), as discussed in the preceding section, which results in an

2.2 General critical points

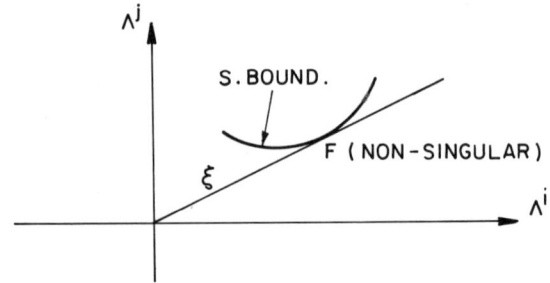

Figure 2.16

asymmetric point of bifurcation, and the general points obtained this way are associated with the undeflected configuration of the system. It follows that the rays determined by (2.2.79) and (2.2.80) yield the tangent lines to the equilibrium surface (and consequently to the stability boundary) as well as the singular critical points.

Figures 2.16 and 2.17 illustrate these two phenomena respectively.

On the basis of the foregoing discussion the following theorem can be stated:

THEOREM 2.5. *Singular critical points associated with a one-degree-of-freedom system lie on planes in the load-space.*

COROLLARY 1. *If $M=2$ there can be only one singular critical point.*

COROLLARY 2. *If $M=3$, singular critical points lie on a straight line in the load-space.*

It is understood that while Theorem 2.5 is valid for the entire load-space, Theorem 2.4 is an asymptotic result.

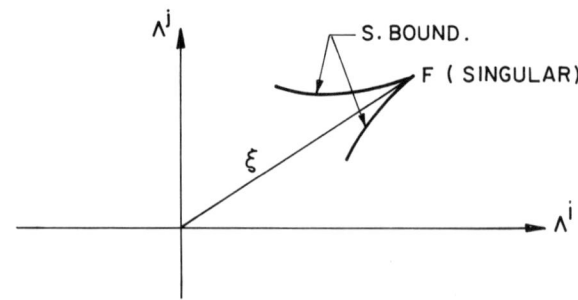

Figure 2.17

2.2.7 Stability distribution on the equilibrium surface

Locating singular critical points associated with N-degree-of-freedom systems may present difficulties unless the buckling modes associated with bifurcations can *a priori* be predicted. However, in many practical problems, a singular point will normally arise when a particular ray $\Lambda^i = l^i \xi$ is followed along which the structure remains undeflected or practically so, and the buckling mode may be known. The same principles as in the analysis of one-degree-of-freedom systems can, then, be applied in conjunction with (2.2.7) and (2.2.15) or (2.2.16). Thus, the equations

$$Q_1^k \Lambda^k = 0$$
$$A_0 + A^k \Lambda^k = 0$$

yield singular points where Q_1 is assumed to be the buckling mode.

2.2.7 Stability distribution on the equilibrium surface

In order to examine the stability of the equilibrium states in the vicinity of F, it will be assumed that F is *general primary* so that $\Pi_{11} = 0$, $\Pi_{ss} > 0$ for $s \neq 1$. It is recalled that for the stability of a noncritical equilibrium state the second variation of the energy function must be positive definite. This implies that the stability determinant

$$\Delta(u_k, \varphi^i) \equiv |\Pi_{ij}(u_k, \varphi^i)| \qquad (2.2.82)$$

evaluated at that state must be positive. If the stability determinant is negative, then, the equilibrium state is unstable. It must be noted, however, that the assessment of stability by examining the sign of the stability determinant is only valid in the vicinity of a primary critical point since the stability coefficients cannot change sign in this small region before passing through zero.

In order to evaluate the stability determinant at an equilibrium state on the equilibrium surface in the vicinity of F, consider the expansion of this determinant around F,

$$\Delta = \Delta_i u_i + \Delta^i \varphi^i$$
$$+ \frac{1}{2!}(\Delta_{ij} u_i u_j + 2\Delta_i^j u_i \varphi^j + \Delta^{ij} \varphi^i \varphi^j)$$
$$+ \frac{1}{3!}(\cdots) + \cdots \qquad (2.2.83)$$

Substituting for Δ_i and Δ^i from (2.2.46), one has

$$\Delta = (\Pi_{11i} u_i + \Pi_{11}^i \varphi^i) \prod_{s=2}^{N} \Pi_{ss} + \frac{1}{2!}(\cdots) + \cdots \qquad (2.2.84)$$

2.2 General critical points

Evaluating (2.2.84) at an arbitrary equilibrium state which is defined by equations (2.2.33) and (2.2.34) and keeping to a first order approximation result in

$$\Delta = (\Pi_{111}u_1 + c^m\varphi^m)\prod_{s=2}^{N}\Pi_{ss} \tag{2.2.85}$$

It is understood that singular points are not considered here; using the equilibrium surface (2.2.44), however, such points can similarly be treated.

Since attention is being restricted to the neighbourhood of a primary critical point, one has the following stability criterion:

$$\Pi_{111}u_1 + c^m\varphi^m \begin{array}{ll} >0 & \text{stable} \\ =0 & \text{for critical equilibrium} \\ <0 & \text{unstable} \end{array} \tag{2.2.86}$$

It can now be shown that the equilibrium surface is divided by the critical zone into stable and unstable domains.

Consider a critical point A on the critical zone satisfying the equations (2.2.53) to (2.2.55). Neighbouring equilibrium states can be obtained by keeping $\overset{*}{\varphi}{}^m_A$ ($m \neq 1$) constant and giving a small increment τ to $\overset{*}{\varphi}{}^1_A$. Thus, for certain values of τ, the set of

$$\begin{aligned}\varphi^m &= \overset{*}{\varphi}{}^m_A \\ \varphi^1 &= \overset{*}{\varphi}{}^1_A + \tau\end{aligned} \tag{2.2.87}$$

corresponds to certain equilibrium states on the surface. u_1 was previously determined and is given by the equation (2.2.58) as

$$u_1 = \overset{*}{u}{}^A_1 \pm \frac{1}{\Pi_{111}}(-2\Pi_{111}\Pi^1_1\tau)^{\frac{1}{2}} \tag{2.2.88}$$

Using (2.2.35), (2.2.55) and (2.2.87), u_s can also be determined as

$$u_s = \overset{*}{u}{}^A_s - \frac{1}{\Pi_{ss}}\left(\Pi^1_s - \frac{\Pi_{ss}\Pi^1_1}{\Pi_{111}}\right)\tau \tag{2.2.89}$$

Here we are only interested in real equilibrium states, and it will be assumed that the increment τ is given in the right direction to ensure real solutions.

Evaluating the stability determinant at the particular states defined by (2.2.87) to (2.2.89) results in

$$\Delta = [\Pi_{111}\overset{*}{u}{}^A_1 \pm (-2\Pi_{111}\Pi^1_1\tau)^{\frac{1}{2}} + c^m\overset{*}{\varphi}{}^m_A]\prod_{s=2}^{N}\Pi_{ss} \tag{2.2.90}$$

2.2.7 Stability distribution on the equilibrium surface

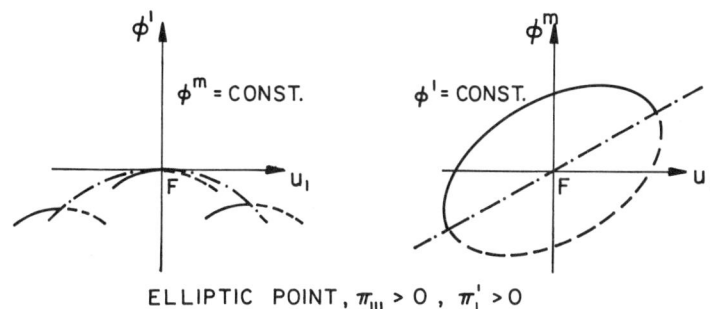

ELLIPTIC POINT, $\pi_{111} > 0$, $\pi'_1 > 0$

Figure 2.18

Using (2.2.54), one has

$$\Delta = \pm(-2\Pi_{111}\Pi^1_1\tau)^{\frac{1}{2}}\prod_{s=2}^{N}\Pi_{ss} \tag{2.2.91}$$

Evidently, the stability determinant (2.2.91) is positive for one solution and negative for the other lying on either side of the critical zone. Moving the point A along the critical zone and assigning certain values to τ, the stability of all equilibrium states in the vicinity of F can thus be examined. The obvious conclusion can be stated as a theorem:

THEOREM 2.6. *The equilibrium surface in the vicinity of a nonsingular general critical point is divided by the critical zone into two distinct domains so that on one side of the critical zone the equilibrium is stable while on the other side the equilibrium is unstable.*

This phenomenon is illustrated in Figures 2.18, 2.19 and 2.20 for elliptic, hyperbolic and parabolic points respectively. The critical zone which takes the form of a line in the pictures is shown as a dash-dotted line.

Finally, it is noted that the critical point F itself (and the points of the critical zone) is unstable if it is not singular and $\Pi_{111} \neq 0$ as in the case of one-parameter systems. Singular critical points of practical interest are also unstable.

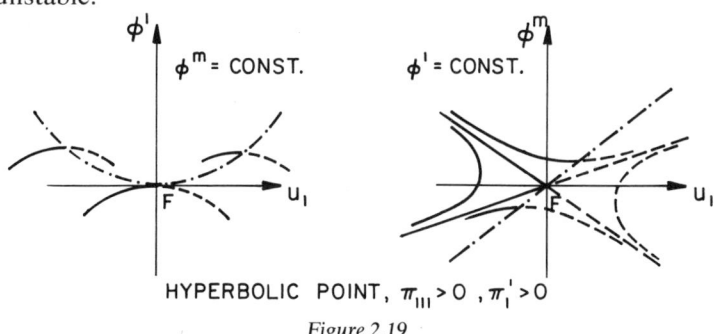

HYPERBOLIC POINT, $\pi_{111} > 0$, $\pi'_1 > 0$

Figure 2.19

2.2 General critical points

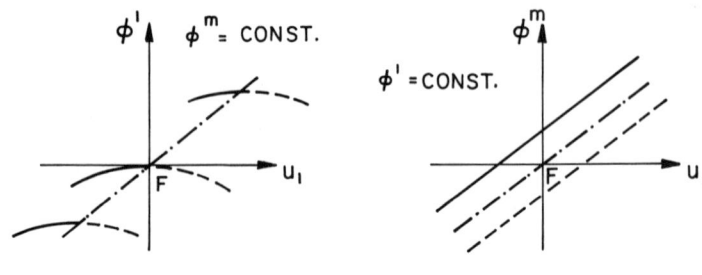

PARABOLIC POINT, $\pi_{III} > 0$, $\pi_I' > 0$

Figure 2.20

2.2.8 Examples, experimental results and discussion

Example 1

The rigid link model used to illustrate limit points in Section 1.4.2 will now be analyzed under combined loading to illustrate *general* critical points.

The loading on the one-degree-of-freedom model structure now consists of a vertical concentrated load, P, at the centre and two equal moments, M, acting at the supports as shown in Figure 2.21a.

The potential energy of the system can be written in the nondimensional form

$$V = \frac{1}{\cos \alpha}\left(1 - \frac{\cos \alpha}{\cos Q_1}\right)^2 - \Lambda^1 \frac{\sin \alpha - \cos \alpha \, \mathrm{tg}\, Q_1}{\cos \alpha}$$

$$- 2\Lambda^2 \frac{\alpha - Q_1}{\cos \alpha} \tag{2.2.92}$$

after it is divided by $kl^2 \cos \alpha$. Here

$$\Lambda^1 = \frac{P}{kl}, \quad \Lambda^2 = \frac{M}{kl^2}$$

Assuming further that the angles α and Q_1 are small, permitting the use of the series expansions

$$(\cos Q_1)^{-1} = 1 + \tfrac{1}{2}Q_1^2 + \tfrac{5}{24}Q_1^4 + \cdots$$

$$\mathrm{tg}\, Q_1 = Q_1 + \tfrac{1}{3}Q_1^3 + \tfrac{2}{15}Q_1^5 + \cdots \tag{2.2.93}$$

$$\cos \alpha = 1 - \frac{\alpha^2}{2!} + \frac{\alpha^4}{4!} + \cdots,$$

ignoring the irrelevant terms (such as constants, etc.), and keeping terms

2.2.8 Examples, experimental results and discussion

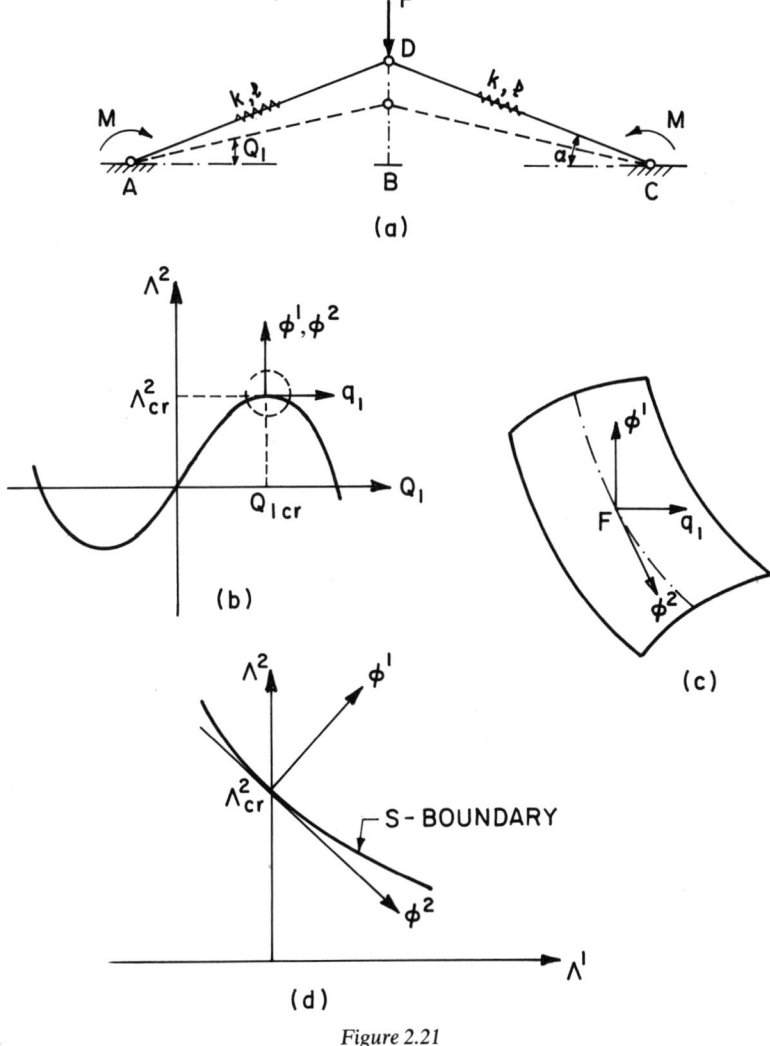

Figure 2.21

only up to the fourth order in Q_1 and α, one finally has

$$V = \tfrac{1}{4}(Q_1^4 - 2\alpha^2 Q_1^2) + \Lambda^1(Q_1 + \tfrac{1}{3}Q_1^3) + 2\Lambda^2 \frac{Q_1}{\cos \alpha} \qquad (2.2.94)$$

This energy function will be used to explore the neighbourhood of the critical point given by

$$\Lambda^1 = 0, \qquad \Lambda^2 = \Lambda_{cr}^2, \qquad Q_1 = Q_{1cr} \qquad (2.2.95)$$

2.2 General critical points

in which case the critical values of Q_1 and Λ^2 can immediately be determined. Thus, differentiating (2.2.94) with respect to Q_1 yields the equilibrium surface

$$Q_1^3 - \alpha^2 Q_1 + \frac{2}{\cos \alpha} \Lambda^2 = 0 \tag{2.2.96}$$

A second differentiation results in

$$3Q_1^2 - \alpha^2 = 0 \tag{2.2.97}$$

The simultaneous solution of (2.2.96) and (2.2.97) gives

$$Q_{1cr} = \pm \frac{\sqrt{3}}{3} \alpha, \qquad \Lambda_{cr}^2 = \pm \frac{\sqrt{3}}{9} \alpha^3 \cos \alpha \tag{2.2.98}$$

Shifting the coordinates to this critical point by the transformation of translation

$$\begin{aligned} Q_1 &= +Q_{1cr} + q_1 \\ \Lambda^2 &= +\Lambda_{cr}^2 + \lambda^2 \\ \Lambda^1 &= 0 + \lambda^1 \end{aligned} \tag{2.2.99}$$

and ignoring irrelevant terms one has

$$V = \tfrac{1}{4}(4Q_{1cr}q_1^3 + q_1^4) + \lambda^1[(1+Q_{1cr}^2)q_1 + Q_{1cr}q_1^2 + \cdots]$$

$$+ 2\lambda^2 \frac{q_1}{\cos \alpha} \tag{2.2.100}$$

where q_1, λ^1, λ^2 are small increments in the corresponding variables.
Introduce now the transformation of rotation

$$\begin{aligned} \lambda^1 &= \frac{B}{(1+B^2)^{\frac{1}{2}}} \varphi^1 + \frac{1}{(1+B^2)^{\frac{1}{2}}} \varphi^2 \\ \lambda^2 &= \frac{1}{(1+B^2)^{\frac{1}{2}}} \varphi^1 - \frac{1}{(1+B^2)^{\frac{1}{2}}} \varphi^2 \end{aligned} \tag{2.2.101}$$

to obtain a canonical representation of the linear form

$$(1+Q_{1cr}^2)\lambda^1 + \frac{2}{\cos \alpha} \lambda^2,$$

where

$$B = \tfrac{1}{2}(1+Q_{1cr}^2) \cos \alpha.$$

2.2.8 Examples, experimental results and discussion

Substituting for λ^1 and λ^2 in the equation (2.2.100), one gets

$$\Pi(u_1, \varphi^i) \equiv V(u_1, \alpha^{ij}\varphi^j) = \tfrac{1}{4}(4Q_{1cr}u_1^3 + u_1^4) + \frac{Q_{1cr}}{(1+B^2)^{\frac{1}{2}}} u_1^2\varphi^2$$

$$+ \frac{2(1+B^2)^{\frac{1}{2}}}{\cos \alpha} u_1\varphi^1 + \cdots \qquad (2.2.102)$$

where $u_1 \equiv q_1$ and α^{ij} is the transformation matrix in (2.2.101).
The first order equilibrium equation is now in the form

$$3Q_{1cr}u_1^2 + 2\frac{Q_{1cr}}{(1+B^2)^{\frac{1}{2}}} u_1\varphi^2 + 2\frac{(1+B^2)^{\frac{1}{2}}}{\cos \alpha} \varphi^1 = 0 \qquad (2.2.103)$$

which defines an *anticlastic* equilibrium surface in the vicinity of the critical point (2.2.95) in the three dimensional load-deflection space (Figures 2.21b and 2.21c).

In order to determine the stability boundary, differentiate (2.2.102) with respect to u_1 for a second time to get

$$6Q_{1cr}u_1 + 2\frac{Q_{1cr}}{(1+B^2)^{\frac{1}{2}}} \varphi^2 = 0 \qquad (2.2.104)$$

Substituting for u_1 in the equation (2.2.103) results in the stability boundary in the vicinity of the critical point (2.2.95),

$$2\frac{(1+B^2)^{\frac{1}{2}}}{\cos \alpha} \overset{*}{\varphi}{}^1 - \frac{1}{3}\frac{Q_{1cr}}{(1+B^2)} (\overset{*}{\varphi}{}^2)^2 = 0 \qquad (2.2.105)$$

Comparing the signs of the curvatures

$$\overset{*}{\varphi}{}^{1,22} = \frac{1}{3} \frac{Q_{1cr} \cos \alpha}{(1+B^2)^{\frac{3}{2}}}$$

and $\qquad (2.2.106)$

$$\varphi^1_{11} = -3 \frac{Q_{1cr} \cos \alpha}{(1+B^2)^{\frac{1}{2}}}$$

one observes that they have opposite signs indicating that the stability boundary is, indeed, convex towards the region of existence in compliance with Theorem 2.3 (Figure 2.21d).

Example 2

This example is concerned with locating singular critical points. Consider the one-degree-of-freedom model consisting of three pin-jointed rigid links of length L as shown in Figure 2.22, and having an initial position prescribed by α_0. Each of the two supports, a and b, has a linear torsional spring of stiffness k. Two concentrated loads, Λ^1 and Λ^2, act

2.2 General critical points

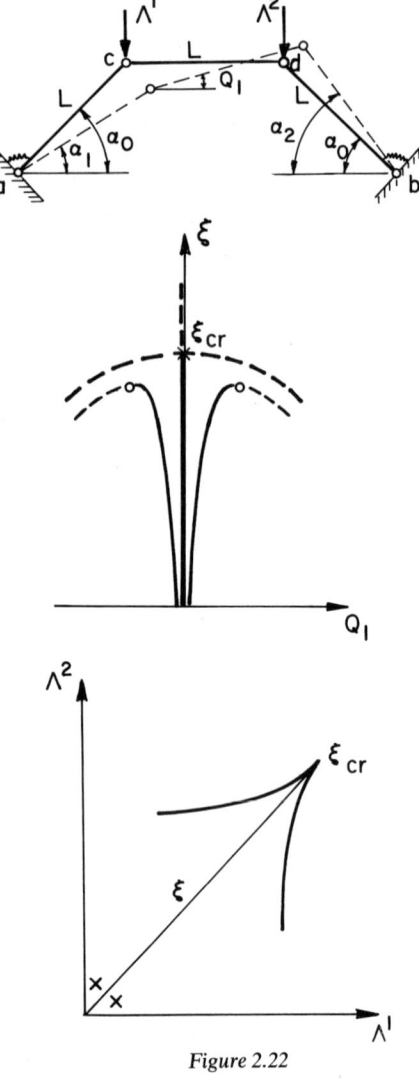

Figure 2.22

vertically at the joints c and d respectively. Choosing the rotation of the member cd as the generalized coordinate Q_1, one has the following compatibility equations

$$\cos \alpha_1 + \cos Q_1 + \cos \alpha_2 = 2 \cos \alpha_0 + 1$$
$$\sin \alpha_1 + \sin Q_1 - \sin \alpha_2 = 0$$
(2.2.107)

where $\alpha_1 = \alpha_1(Q_1)$ and $\alpha_2 = \alpha_2(Q_1)$.

2.2.8 Examples, experimental results and discussion

By successive differentiation and evaluation at $Q_1 = 0$, these equations yield the derivatives

$$\frac{d\alpha_1}{dQ_1}\bigg|_{Q_1=0} = -\frac{d\alpha_2}{dQ_1}\bigg|_{Q_1=0} = -\frac{1}{2\cos\alpha_0}$$

$$\frac{d^2\alpha_1}{dQ_1^2}\bigg|_{Q_1=0} = \frac{d^2\alpha_2}{dQ_1^2}\bigg|_{Q_1=0} = -\frac{1}{2}\left(\frac{1}{\sin\alpha_0} + \frac{1}{\sin 2\alpha_0}\right) \quad (2.2.108)$$

$$\frac{d^3\alpha_1}{dQ_1^3}\bigg|_{Q_1=0} = -\frac{d^3\alpha_2}{dQ_1^3}\bigg|_{Q_1=0} = \cdots$$

which will be used in what follows.

The total potential energy function of the system is in the form

$$V = \tfrac{1}{2}k(\alpha_1 - \alpha_0)^2 + \tfrac{1}{2}k(\alpha_2 - \alpha_0)^2 + \Lambda^1 L \sin\alpha_1$$
$$+ \Lambda^2 L \sin\alpha_2 \quad (2.2.109)$$

To locate the singular critical point, one considers a certain combination of loads, described by $\Lambda^i = l^i \xi$ which satisfies the equations (2.2.79) and (2.2.80). Thus, evaluating the derivatives V_1^1 and V_1^2 from the energy function (2.2.109) and substituting into (2.2.79) yield

$$\frac{d\alpha_1}{dQ_1} l^1 + \frac{d\alpha_2}{dQ_1} l^2 = 0$$

which, by virtue of (2.2.108), results in

$$l^1 = l^2 \quad (2.2.110)$$

as expected.

Differentiating the function (2.2.109) twice with respect to Q_1, evaluating at $Q_1 = 0$ and introducing the resulting derivatives into (2.2.80) together with (2.2.110) yield

$$\xi_{cr} = \frac{\sqrt{2}}{4} \frac{k}{Lf \cos^2\alpha_0} \quad (2.2.111)$$

where

$$f = \frac{1 + 2\cos^3\alpha_0}{2\sin\alpha_0 \cos^2\alpha_0}$$

To establish whether or not the critical point determined by equations (2.2.110) and (2.2.111) is singular, differentiate the energy function (2.2.109) for a third time with respect to Q_1 to obtain

$$\frac{\partial^3 V}{\partial Q_1^3}\bigg|_{Q_1=0} = 0 \quad (2.2.112)$$

153

2.2 General critical points

which indicates that the critical point is singular as defined in the theory. In other words, by increasing the loads Λ^1 and Λ^2 along the ray defined by (2.2.110), an unstable symmetric critical point given by (2.2.111) is reached (provided α_0 is chosen appropriately); it can be shown, however, that a deviation from this ray leads to limit points as shown in Figure 2.22, and the stability boundary takes the form of a sharp cusp in the vicinity of the critical point. Again, if α_0 is not appropriate, the model may not exhibit instability under one of the loads only, or even when the combination of loads is not appropriate, i.e., the stability boundary may exhibit *cut-off* values associated with continuously rising equilibrium paths.

Problem. Assuming that $\alpha_0 = \dfrac{\Pi}{4}$ and $l^1 = l^2$ determine the curvature $d^2\xi/dQ_1^2|_{\xi=\xi_{cr}}$ of the symmetric post-buckling path and show that the critical point is unstable. Consider a slightly different ray and show that the corresponding critical value of ξ is less than the critical value corresponding to $l^1 = l^2$.

Experimental results

The experiments reported here were performed in the Structural Mechanics Laboratory of the Department of Civil Engineering, University College, London [43].

(i) Stability of a shallow frame

A series of experiments has been carried out on a simply supported shallow frame [43] with a 24-inch span which was designed by Roorda [36] for other purposes. It was built up of two identical high tensile steel members with dimensions 1 inch by 0·04 inch which were joined rigidly with a contained angle of 160°. Two loads were applied vertically as shown in Figure 2.23a. P_1 was applied first and kept constant while P_2 was increased gradually until snap-through occurred on reaching a limit point after large deflections. This was repeated for different but constant values of P_1. Then, in order to have a check, P_2 was applied first and kept constant while P_1 was increased until snap-through occurred. The critical values of P_1 and P_2, nondimensionalized by dividing by P_{01} and P_{02} respectively, were plotted to give the stability boundary of the system as shown in Figure 2.23b. P_{01} and P_{02} are the critical values of P_1 and P_2 respectively obtained by applying each load individually.

The overall buckling behaviour of this frame can be approximated to that of a one-degree-of-freedom system by choosing the rotation of the

2.2.8 Examples, experimental results and discussion

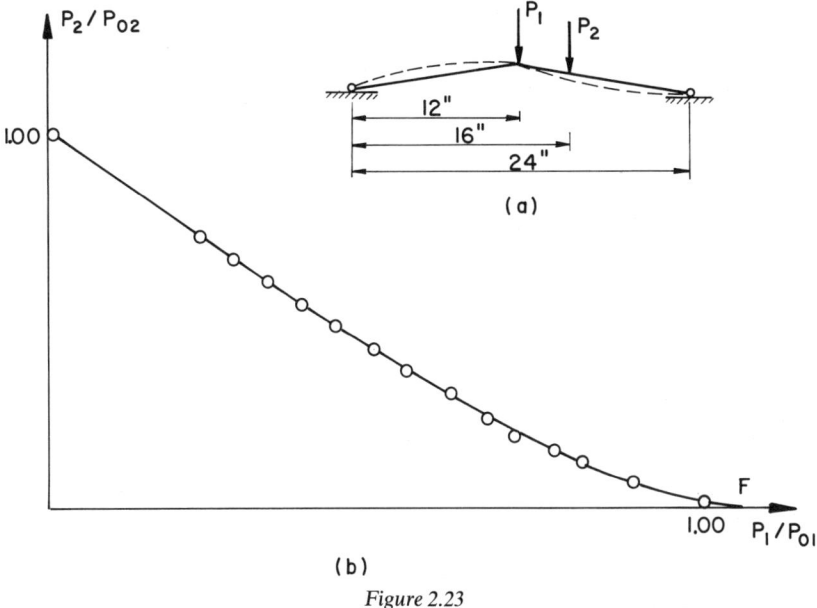

Figure 2.23

crown as a generalized coordinate. Every point on the stability boundary, except F, is associated with a *general* critical equilibrium state. F corresponds to a *singular* general critical state in the overall two-parameter picture, and appears as an unstable symmetric point on a plot of $P_1 - Q_1$, where Q_1 is the rotation of the crown. Obviously, the stability boundary exhibits the features predicted by the general theory, and it is convex towards the region of existence, thus obeying Theorem 2.3.

Similar tests were performed after applying the loads at certain different points. The resulting interaction curves exhibit the same basic property concerning the convexity of the stability boundary as Theorem 2.3 asserts.

(ii) Stability of a shallow arch

A series of experiments on simply supported shallow arches made of high tensile steel strips with dimensions 1 inch by $\frac{1}{16}$ inch (or $\frac{1}{32}$ inch) has been carried out [43]. The strips were rolled to form an approximately sinusoidal curve with a span of 24 inches and various rises. The test reported here was made on an arch with a rise of 1·30 inch. Two loads were applied vertically as illustrated in Figure 2.24a. P_1 was applied first and kept constant while P_2 was increased gradually until snap-through occurred after large deflections. This was repeated for different but

2.2 General critical points

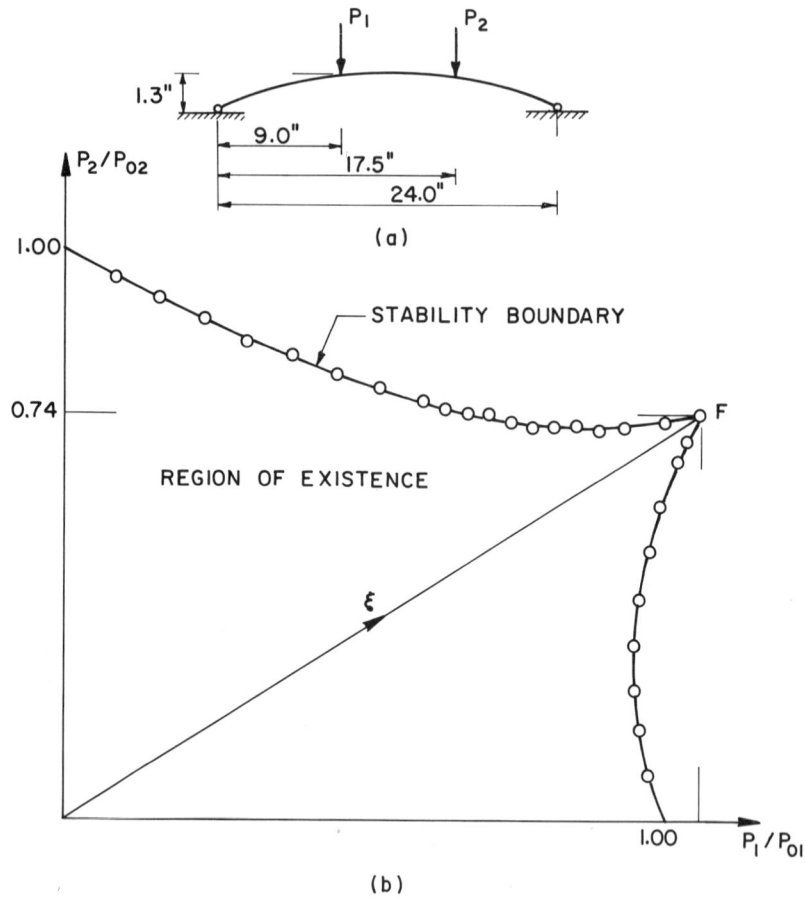

Figure 2.24

constant values of P_1. Then, P_2 was applied first and held constant while P_1 was increased gradually until snap-through occurred. Combinations of critical loads thus obtained are plotted as shown in Figure 2.24b to give the stability boundary of the system. P_{01} and P_{02} are again the critical values of P_1 and P_2 respectively which arise when these loads are applied individually.

The overall buckling behaviour of this arch can be assumed to be described approximately by the rotation of the crown as in the shallow frame, thus reducing the system to a one-degree-of-freedom model. It is seen that the stability boundary consists of two continuous strictly convex curves joining at a singular critical point F. Each and every point

2.2.8 Examples, experimental results and discussion

on both branches, except F, is associated with a *general* critical equilibrium state, and obviously both the branches are convex towards the region of existence. In the vicinity of F the curves take the form of a cusp. It has been observed during the test that the arch remained practically undeflected as the loading was increased in the OF direction, and it swayed to the right or left side equally well when the point F was reached.

Similar tests on arches with different thicknesses and rises, and under the combined action of two loads applied at various points (negative loads were also applied) yield similar curves which are all in compliance with the general theory.

Discussion

Experimental results reported here indicate that the buckling behaviour of continuous structures like shallow arches and frames under combined loading is analogous to that of a one-degree-of-freedom system, thus complying with Theorem 2.3. A survey of the literature shows that the general critical points arising in the stability analysis of particular problems in this category are, indeed, mostly *hyperbolic*, and fall, therefore, within the scope of Theorem 2.3, or in a broader sense Corollary 2 of Theorem 2.1. Examples for the type of behaviour described by Theorem 2.1, however, are hard to find. The investigation of Loo and Evan-Ivanowski [86] on the buckling behaviour of a shallow spherical shell under the action of combined external pressure and concentrated load at the apex seems to be the only significant problem satisfying this theorem. The loss of stability of this system is always associated with *elliptic* general critical points at which bifurcation is ruled out, and the stability boundary is indeed *concave* towards the origin as stated in Theorem 2.1.

Despite this example, the majority of structural systems whose loss of stability is associated with general critical points exhibits convex (towards the origin) stability boundaries in direct contrast with linear bifurcating systems such as plates under combined loading (see Section 2.4.3). The practical significance of this property lies in the fact that a straight line joining two points on the boundary yields an upper bound rather than a lower bound for systems obeying Theorem 2.1. Consider, for example, Figure 2.23; if the points on the axes are joined the segment of the straight line between these points provides an upper bound for the stability boundary, and if Papkovich's Theorem, which is valid for linear bifurcating cases only, was applied, this straight line would have been regarded as a lower bound leading to a totally unsafe erroneous estimate.

2.2 General critical points

It can be shown [51], that if two independent loads can be combined linearly, such that an increase in the loading along this particular combination (i.e., along a particular ray $\Lambda^i = l^i \xi$) leads to a bifurcation point, the stability boundary of that system is convex towards the origin. If the bifurcation point arising along the loading ray is unstable symmetric, then, that point is a *singular* critical point in the overall picture, and the stability boundary takes the form of a cusp in its vicinity as shown in Figure 2.24. Although the cusp is not apparent in Figure 2.23, the point F is also a singular point and the cusp would have appeared fully if negative values of P_2 were considered. Obviously, the ray leading to the singular point in this case is defined by $l^1 = 1$ and $l^2 = 0$, i.e., the loading must be along P_1.

Estimates of the stability boundary are, of course, affected by the presence of a singular critical point. If the critical points on the P_1 and P_2 axes in Figure 2.24 are joined, for example, the straight line segment between these points now represents a lower bound. This is due to the fact that Theorem 2.3 and/or Corollary 2 of Theorem 2.1 are formulated for nonsingular points and they can only be applied to the branches of the stability boundary on each side of F separately. It seems, therefore, that the location of singular points must be known in order to be able to obtain upper and/or lower bound estimates of the stability boundary on the basis of the convexity properties. As remarked in Section 2.2.6, however, a linear eigenvalue analysis will often yield the singular points readily, and once the singular points are located, the theorems can be applied safely to obtain upper and/or lower bounds. By joining F to the critical points on the axes (in Figure 2.24), for example, one establishes upper bounds. The analysis leading to the critical points on the axes is relatively much simpler than obtaining other points on the boundary since the interaction of loads is not then involved. It is even possible that the critical value of each load applied individually is available in the literature, and upper/lower bounds can be established without much analytical effort.

2.3

Special critical points: asymmetric

In the theory of elastic stability there exists a considerable number of buckling problems in which only *special* critical points arise, all possible one-parameter combinations of the Λ^i leading to a bifurcation of the solution. Frames and plates under the action of certain combinations of axial compression, shear, etc., for instance, lose stability always at a special critical point. Generally, symmetric perfect systems may be expected to exhibit special critical points.

In Chapter 2.1, a special point was characterized by $H_{11} = 0$, $H_{ss} \neq 0$ ($s \neq 1$) and $\overrightarrow{\text{grad}}_\lambda H_1(0,0) = 0$ which rules out the possibility of an extremum on the equilibrium surface. In fact, it was seen that the equilibrium surface in the vicinity of such a critical point is an *improper* or *degenerate* one, and in practical situations often takes the form of two intersecting multi-dimensional surfaces (hyper-surfaces) or planes. It is this property of the equilibrium surface that allows for the introduction of certain appropriate transformations for a more detailed investigation of the post-critical behaviour and the stability boundary in the vicinity of a special critical point. Symmetric systems will be treated separately in Chapter 2.4.

2.3.1 Equilibrium surface in the vicinity of a special critical point

Consider the systems described by the potential energy function (2.1.1) and suppose the equilibrium equations $V_i(Q_j, \Lambda^k) = 0$ are solved simultaneously to yield results in the form $Q_i = Q_i(\Lambda^j)$ which define certain equilibrium surfaces in the $M + N$ dimensional $Q_i - \Lambda^j$ space. Consider the particular surface

$$Q_i = Q_i^F(\Lambda^j) \qquad (2.3.1)$$

which passes through the origin representing the unbuckled solution. In analogy with one-parameter systems, this surface will be called *fundamental*. It is assumed that the fundamental surface (2.3.1) is single-valued

2.3 Special critical points: asymmetric

at least in the region of interest. That is to say, there is a one-to-one correspondence between a set of Λ^j and Q_i^F, thus excluding the possibility of an extremum and consequently of a *general* critical point on the surface. The potential energy can, then, be referred to the fundamental surface by introducing the sliding coordinates q_i and setting

$$Q_i = Q_i^F(\Lambda^j) + q_i \qquad (2.3.2)$$

Here, the q_i represent a small increment *from the fundamental surface* (Figure 2.25). A further change of coordinates by means of a linear orthogonal transformation,

$$q_i = \alpha_{ij}(\Lambda^l)u_j, \qquad \alpha_{ij}(\Lambda^l)\alpha_{jk}(\Lambda^l) = \delta_{ik}, \qquad (2.3.3)$$

is introduced to diagonalize the quadratic form of the energy function at each and every point of the fundamental surface in the region of interest.

Introducing (2.3.2) and (2.3.3) into the energy function, one gets the new function

$$S(u_i, \Lambda^k) \equiv V[Q_i^F(\Lambda^k) + \alpha_{ij}(\Lambda^k)u_j, \Lambda^k] \qquad (2.3.4)$$

with the properties

$$\begin{aligned}S_i(0, \Lambda^j) &= S_i^j(0, \Lambda^k) = S_i^{jk}(0, \Lambda^l) = \cdots = 0\\ S_{ij}(0, \Lambda^k) &= S_{ij}^k(0, \Lambda^l) = S_{ij}^{kl}(0, \Lambda^m) = \cdots = 0 \quad \text{for} \quad i \neq j\end{aligned} \qquad (2.3.5)$$

in the self-evident notation. The first of these properties follow immediately from the fact that $u_i = 0$ defines the fundamental surface, and

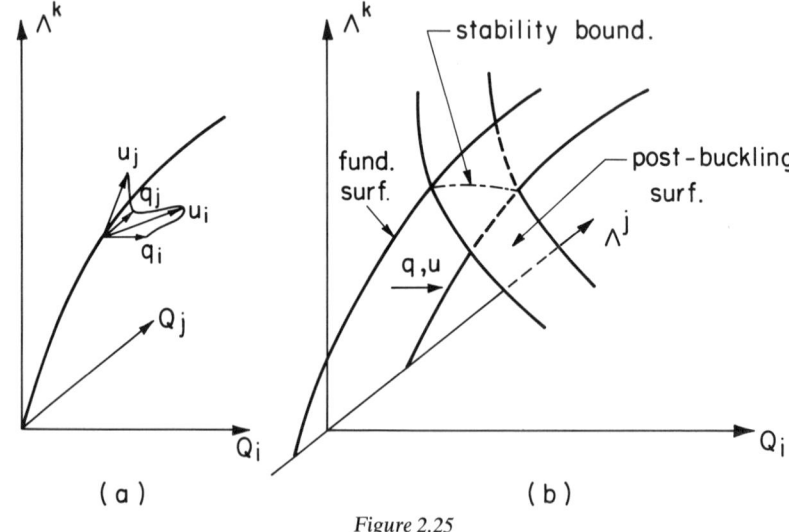

Figure 2.25

2.3.1 Equilibrium surface in the vicinity of a special critical point

the second is due to the transformation (2.3.3) which is a function of the parameters Λ^i so that the quadratic form of the energy function is diagonalized at each and every point on the surface. In other words, a given set of Λ^i in the region of interest defines an equilibrium state (Λ_F^i, Q_i^F) on the fundamental surface and an orthogonal transformation matrix $\alpha_{ij}(\Lambda^k)$ which diagonalizes the quadratic form of the energy at that point; as the parameters Λ^i vary, the u_i axes slide on the fundamental surface and rotate, thus keeping the quadratic form diagonalized at every point on the surface.

A transformation of the Λ^i will, finally, be introduced to achieve a *canonical* representation of the linear form corresponding to $S_{11}^i(0, \overset{*}{\Lambda}{}_F^j)$ in which $\overset{*}{\Lambda}{}_F^i$ are the critical values of Λ^i at the critical point F where $S_{11} = 0$. Thus, the linear orthogonal transformation

$$\Lambda^i = \gamma^{ij} \Phi^j \tag{2.3.6}$$

is chosen such that when (2.3.6) is substituted in the energy function $S(u_i, \Lambda^j)$ the resulting function

$$R(u_i, \Phi^j) \equiv S(u_i, \gamma^{ij}\Phi^j) \tag{2.3.7}$$

has the properties

$$R_{11}^i(0, \overset{*}{\Phi}{}_F^i) \neq 0, \quad \text{all} \quad R_{11}^m(0, \overset{*}{\Phi}{}_F^i) = 0 \tag{2.3.8}$$

where $m \neq 1$, and the $\overset{*}{\Phi}{}_F^i$ are the critical values of Φ^i at F, the point of interest.

It can be shown that the properties (2.3.5) are now replaced by

$$\begin{aligned} R_i(0, \Phi^j) &= R_i^j(0, \Phi^k) = \cdots = 0 \\ R_{ij}(0, \Phi^k) &= R_{ij}^k(0, \Phi^l) = \cdots = 0 \quad \text{for} \quad i \neq j \end{aligned} \tag{2.3.9}$$

This new function will be used to explore the post-critical behaviour of the system in the vicinity of the special critical point F at which

$$R_{11}(0, \overset{*}{\Phi}{}_F^i) = 0 \quad \text{and all} \quad R_{ss}(0, \overset{*}{\Phi}{}_F^i) \neq 0 \quad (s \neq 1) \tag{2.3.10}$$

and it is understood that the only necessary properties of this function are given by (2.3.8) to (2.3.10).

In analogy with Section 2.2.2, one can now choose u_1 and Φ^m ($m \neq 1$) as the independent parameters in the beginning of the analysis, and express *the post-buckling equilibrium surface* as

$$u_s = u_s(u_1, \Phi^m) \qquad \Phi^l = \Phi^l(u_1, \Phi^m) \tag{2.3.11}$$

If these functions are substituted back into the equilibrium equations $R_i = 0$, one has the identities

$$R_i[u_j(u_1, \Phi^m), \Phi^k(u_1, \Phi^m)] \equiv 0 \tag{2.3.12}$$

2.3 Special critical points: asymmetric

Differentiating (2.3.12) once with respect to u_1 and once with respect to Φ^m ($m = 2, 3, \ldots, M$) yields

$$R_{ij}u_{j,1} + R_i^j\Phi_1^j = 0$$

and (2.3.13)

$$R_{ij}u_j^m + R_i^j\Phi^{j,m} = 0$$

which upon evaluation at F where $u_i = 0$ and $\Phi^j = \overset{*}{\Phi}{}^j_F$ result in [1]

$$u_{s,1} = 0, \quad u_s^m = 0 \quad \text{for} \quad i = s \neq 1 \tag{2.3.14}$$

while the equations are identically satisfied for $i = 1$.

Differentiating the first equation in (2.3.13) with respect to u_1 for a second time, one has

$$(R_{ijk}u_{k,1} + R_{ij}^k\Phi_1^k)u_{j,1} + R_{ij}u_{j,11}$$
$$+ (R_{ik}^j u_{k,1} + R_i^{jk}\Phi_1^k)\Phi_1^j + R_i^j\Phi_{11}^j = 0$$

which yields

$$\Phi_1^1 = -\frac{R_{111}}{2R_{11}^1} \quad \text{for} \quad i = 1$$

and (2.3.15)

$$u_{s,11} = -\frac{R_{s11}}{R_{ss}} \quad \text{for} \quad i = s$$

Differentiate now the second equation in (2.3.13) with respect to Φ^n ($n = 2, 3, \ldots, M$) to get

$$(R_{ijk}u_k^n + R_{ij}^k\Phi^{k,n})u_j^m + R_{ij}u_j^{mn}$$
$$+ (R_{ik}^j u_k^n + R_i^{jk}\Phi^{k,n})\Phi^{j,m} + R_i^j\Phi^{j,mn} = 0 \tag{2.3.16}$$

giving on evaluation

$$u_s^{mn} = 0 \tag{2.3.17}$$

Differentiating the first equation in (2.3.13) with respect to Φ^m ($m = 2, 3, \ldots, M$) yields

$$(R_{ijk}u_k^m + R_{ij}^k\Phi^{k,m})u_{j,1} + R_{ij}u_{j,1}^m$$
$$+ (R_{ik}^j u_k^m + R_i^{jk}\Phi^{k,m})\Phi_1^j + R_i^j\Phi_1^{j,m} = 0$$

---

[1] It is understood that

$$u_{1,1} = 1, \quad u_{1,11} = 0, \text{ etc.}$$
$$\Phi^{1,1} = 1, \quad \Phi^{1,11} = 0, \text{ etc.}$$

2.3.1 Equilibrium surface in the vicinity of a special critical point

which results in

$$\Phi^{1,m} = 0, \quad u_{s,1}^m = 0 \qquad (2.3.18)$$

It seems that for certain second derivatives a third order perturbation is required, and the differentiation of (2.3.16) once with respect to u_1 and once with respect to Φ^r ($r = 2, 3, \ldots, M$) yields upon evaluation at F

$$\Phi^{1,mn} = -\frac{R_{11}^{mn}}{R_{11}^1}, \quad u_{s,1}^{mn} = 0, \quad u_s^{mnr} = 0 \qquad (2.3.19)$$

One now has the required derivatives to construct the asymptotic equation of the post-buckling surface. Thus, using (2.3.14), (2.3.15) and (2.3.17) to (2.3.19) leads to the first order equations

$$2R_{11}^1 \varphi^1 + R_{111} u_1 + R_{11}^{mn} \varphi^m \varphi^n = 0$$

and (2.3.20)

$$2R_{ss} u_s + R_{s11} u_1^2 = 0$$

which define the post-buckling surface in the vicinity of the *special critical point* F, and where $\varphi^i = \Phi^i - \overset{*}{\Phi}{}^i_F$. The equations (2.3.20) were first derived in [43] and later presented in [49].

Various intersections of this surface are shown in Figure 2.26. It is observed that the post-buckling behaviour is in general asymmetric and the special critical point under consideration, therefore, can be called *asymmetric* in analogy with one-parameter systems. Figure 2.26c describes the interesting phenomenon mentioned in Section 1.1.6 (see also Figure 1.5d), and one notes that this type of behaviour is produced when a loading ray in the φ^m-subspace is followed. It will be seen in the next section that $\varphi^1 = 0$ represents the plane which is tangential to the stability boundary at F, and it follows that a loading ray or a path which is tangential to the stability boundary of the system results in two tangential equilibrium paths which *do not exchange* stability (see Section 2.3.3) as indicated in Figure 2.26c which is drawn for $R_{11}^1 < 0$, $R_{111} > 0$ and $R_{11}^{mn} < 0$. It is understood that the path $u_1 = 0$ in this figure is on the fundamental surface, and since the picture is drawn for $R_{11}^1 < 0$ and $R_{11}^{mm} < 0$, this particular fundamental path is unstable. The equation of the post-buckling path tangential to the path $u_1 = 0$ is readily obtained from (2.3.20) as

$$R_{111} u_1 + R_{11}^{mm} (\varphi^m)^2 = 0 \qquad (2.3.21)$$

which is derived for a particular parameter φ^m. Evidently, the transformations introduced in the beginning of this section and, in fact, the whole formulation of the problem contributed towards obtaining the equilibrium surface (2.3.20) and consequently (2.3.21) in the simplest possible forms,

2.3 Special critical points: asymmetric

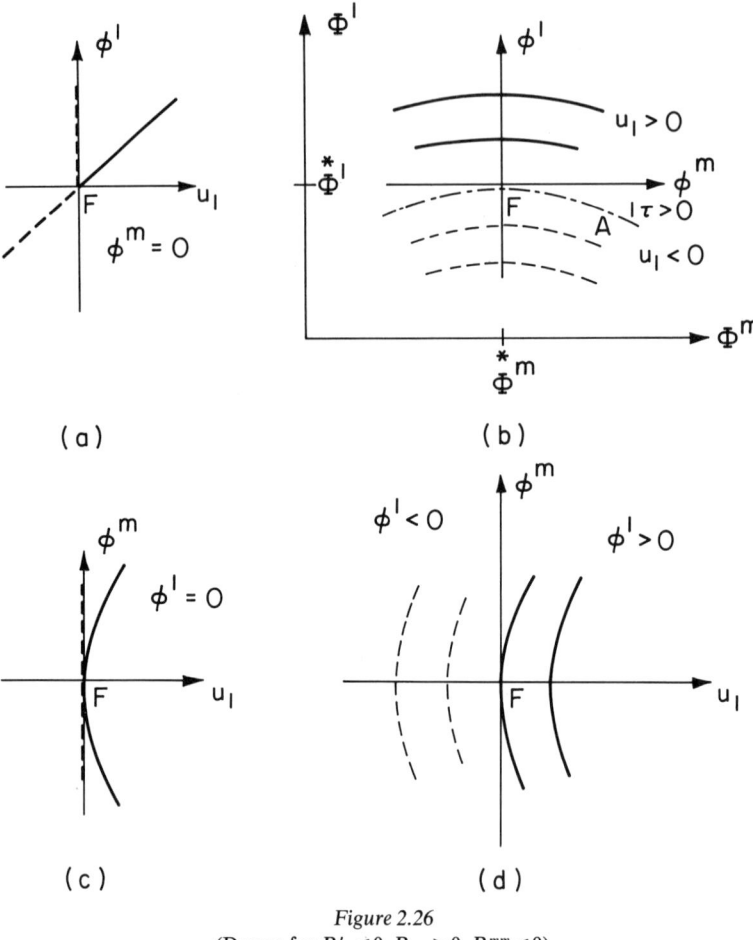

Figure 2.26
(Drawn for $R'_{11}<0$, $R_{111}>0$, $R^{mm}_{11}<0$)

and one immediately observes that the condition ($\varphi^1 = 0$) giving rise to the behaviour described in (2.3.21) and $u_1 = 0$ is equivalent to assuming that the particular coefficient $R^1_{11} = 0$ at F. Thus, generally speaking, for this type of behaviour *the derivative of the vanishing stability coefficient with respect to the loading parameter should also vanish simultaneously.*

2.3.2 Stability boundary

In analogy with one-parameter systems which produce a critical point at the intersection of two distinct paths, it is conceivable that the intersection of two equilibrium surfaces here represents a critical surface, and in

2.3.2 Stability boundary

particular the first intersection of the fundamental surface with a post-buckling surface yields the stability boundary. One should, then, obtain the stability boundary of the system under consideration (in the vicinity of F) by simply setting $u_i = 0$ in (2.3.20). It will now be shown analytically that this is indeed true.

To this end, introduce the stability determinant

$$\Delta(u_k, \varphi^l) \equiv \det |R_{ij}(u_k, \varphi^l)| \qquad (2.3.22)$$

Critical points should satisfy the determinantal equation

$$\Delta(u_k, \varphi^l) = 0$$

Differentiating the determinant (2.3.22) by rows once with respect to u_i and once with respect to φ^1 and evaluating at the critical point F yields

$$\begin{aligned}\Delta_i &= R_{11i} \prod_{s=2}^{N} R_{ss} \\ \Delta^1 &= R_{11}^1 \prod_{s=2}^{N} R_{ss}\end{aligned} \qquad (2.3.23)$$

Differentiating (2.3.22) with respect to φ^m ($m = 2, 3, \ldots, M$) and evaluating at F, one observes that $\Delta^m = 0$. A second differentiation with respect to φ^n ($n = 2, 3, \ldots, M$) yields on evaluation

$$\Delta^{mn} = R_{11}^{mn} \prod_{s=2}^{N} R_{ss} \qquad (2.3.24)$$

Taylor's expansion, then, results in

$$\Delta = (R_{11i}u_i + R_{11}^1\varphi^1 + \tfrac{1}{2}R_{11}^{mn}\varphi^m\varphi^n) \prod_{s=2}^{N} R_{ss} + \cdots \qquad (2.3.25)$$

Evaluating (2.3.25) at an arbitrary state on the fundamental equilibrium surface (i.e., setting $u_i = 0$), one obtains to a first approximation

$$\Delta = (R_{11}^1\varphi^1 + \tfrac{1}{2}R_{11}^{mn}\varphi^m\varphi^n) \prod_{s=2}^{N} R_{ss}$$

which yields the equation of the critical surface (or the stability boundary if all $R_{ss} > 0$ for $s \neq 1$)

$$R_{11}^1 \overset{*}{\varphi}{}^1 + \tfrac{1}{2}R_{11}^{mn} \overset{*}{\varphi}{}^m \overset{*}{\varphi}{}^n = 0 \qquad (2.3.26)$$

in the vicinity of F.

The following theorem can, then, be stated:

THEOREM 2.7. *Primary special critical points lie on the intersection of the fundamental and post-buckling equilibrium surfaces.*

2.3 Special critical points: asymmetric

It is seen that the stability boundary (2.3.26) can be *synclastic* or *nonsynclastic* depending on whether the matrix R_{11}^{mn} is positive (negative) definite or indefinite. If $R_{11}^1 < 0$, the condition ensuring that the stability boundary is convex towards the region of instability is the negative definiteness of the matrix R_{11}^{mn}. Here, the regions of stability and instability are, of course, defined in the load-space with regard to the fundamental equilibrium surface, the former corresponding to the stable domain and the latter to the unstable domain of this surface.

2.3.3 Stability distribution on the equilibrium surfaces

Using the stability determinant (2.3.25) as in Section 2.2.7, one can examine the stability of equilibrium states in the neighbourhood of the primary critical point F at which $R_{11} = 0$, and all $R_{ss} > 0$ for $s \neq 1$.

For this purpose, evaluate the stability determinant (2.3.25) at an arbitrary state on the fundamental surface to get the following stability criterion

$$R_{11}^1 \varphi^1 + \tfrac{1}{2} R_{11}^{mn} \varphi^m \varphi^n \quad \begin{array}{l} > \text{ stable} \\ = 0 \text{ for critical equilibrium} \\ < \text{ unstable} \end{array} \qquad (2.3.27)$$

Considering an arbitrary point A on the stability boundary (2.3.26), one can examine the stability of neighbouring equilibrium states on the basis of this criterion by keeping $\varphi^m = \overset{*}{\phi}{}_A^m = \text{const}$, and giving a small increment τ to φ^1. Thus, for the points defined by

$$\begin{aligned} \varphi^1 &= \overset{*}{\phi}{}_A^1 + \tau \\ \varphi^m &= \overset{*}{\phi}{}_A^m \end{aligned} \qquad (2.3.28)$$

the stability criterion (2.3.27) takes the form

$$R_{11}^1 \tau \quad \begin{array}{l} > \text{ stable} \\ = 0 \text{ for critical equilibrium} \\ < \text{ unstable} \end{array} \qquad (2.3.29)$$

If $R_{11}^1 < 0$, say, then $\tau < 0$ defines *the region of stability*, and $\tau > 0$ the *region of instability* as the point A slides on the stability boundary (Figure 2.26b).

Evidently, the stability boundary divides the fundamental surface into stable and unstable domains. As usual, stable and unstable paths in Figure 2.26 are shown with full and dashed lines, respectively, while a dash-dotted line is used to indicate the stability boundary.

2.3.3 Stability distribution on the equilibrium surfaces

In order to examine the stability of states lying on the post-buckling surface, evaluate the determinant (2.3.25) on this surface which is given by (2.3.20). Thus, substituting for u_s and φ^m into (2.3.25) one has the stability criterion

$\frac{1}{2}R_{111}u_1 = 0$ for critical equilibrium
- $>$ stable
- $<$ unstable

If it is assumed that $R_{111}>0$, then, $u_1>0$ defines the stable post-buckling equilibrium states and $u_1<0$ unstable ones.

Alternatively, substitutions for u_s and u_1 in the determinant (2.3.25) yield upon making use of (2.3.28)

$-R_{11}^1\tau = 0$ for critical equilibrium (2.3.30)
- $>$ stable
- $<$ unstable

Comparing (2.3.29) and (2.3.30), one formulates the following theorem:

THEOREM 2.8. *The intersection of the fundamental and post-buckling equilibrium surfaces, that is the stability boundary, divides both the surfaces into stable and unstable domains so that to the points of the region of stability (region of instability) only unstable (stable) post-buckling states correspond.*

This is analogous to Poincaré's *exchange of stability* concept and can be regarded as a generalization of Poincaré's concept to multiple parameter systems [49].

It is inferred from Theorem 2.8 and the equation of the stability boundary (2.3.26) that if the latter has a negative (positive) curvature in a particular direction φ^m, then, the fundamental path $u_1 = 0$ in Figure 2.26c is totally unstable (stable) while the post-buckling path (2.3.21) is stable (unstable). The latter situation is shown in Figure 2.27.

Thus, one concludes that *the existence of a critical point on an equilibrium path is necessary but not sufficient for a loss of stability of the path*, i.e., a path can continue to be stable (unstable) after passing through a critical point where there exists a second unstable (stable) path passing through the same point and being tangential to the first path.

Finally, it is observed that the *asymmetric special critical point F itself is unstable* due to the finite cubic coefficient R_{111}.

2.3 Special critical points: asymmetric

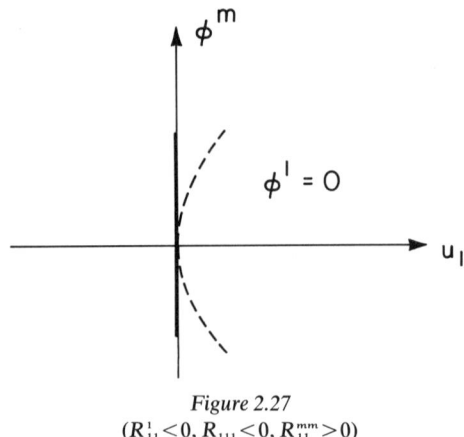

Figure 2.27
($R_{11}^1 < 0, R_{111} < 0, R_{11}^{mm} > 0$)

2.3.4 Linear eigenvalue problems

Although no general theorem concerning the convexity of the stability boundary emerges from the preceding analysis of nonlinear systems, it is possible to establish such a theorem for systems which can be analyzed as a linear multiple-eigenvalue problem.

Consider a system characterized by the total potential energy function

$$V = U_{ij}q_iq_j - \Lambda^k E_{ij}^k q_i q_j \tag{2.3.31}$$

The equilibrium equations

$$(U_{ij} - \Lambda^k E_{ij}^k)q_j = 0 \tag{2.3.32}$$

yield the fundamental equilibrium surface

$$q_i = 0,$$

and the *critical surfaces*

$$\det |U_{ij} - \Lambda^k E_{ij}^k| = 0.$$

The stability boundary is composed of those parts of critical surfaces which are associated with an initial loss of stability and the smallest eigenvalue on each and every loading ray emanating from the origin and intersecting critical surfaces (Figure 2.28).

Consider now an arbitrary *regular* point C on the stability boundary where a stability coefficient (the smallest eigenvalue) vanishes; it can be shown that the equation of the tangent plane at C is given by

$$(U_{ij} - \Lambda^k E_{ij}^k)q_i^c q_j^c = 0 \tag{2.3.33}$$

2.3.4 Linear eigenvalue problems

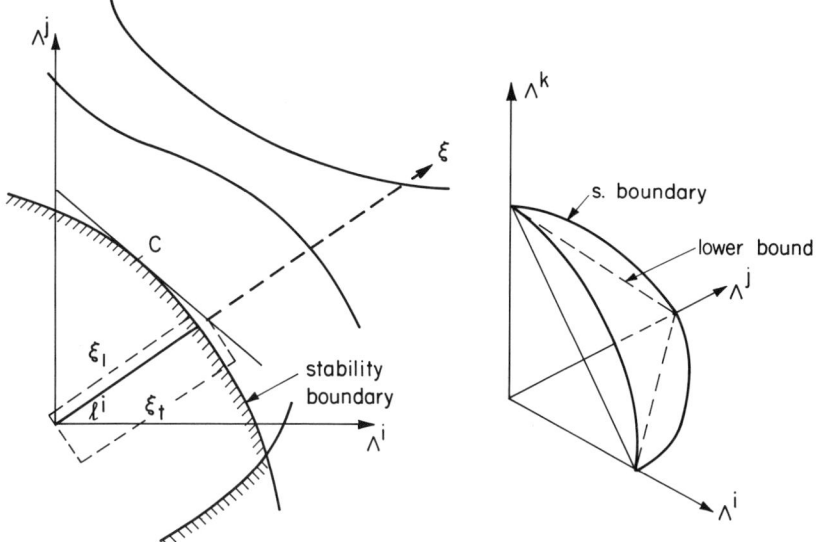

Figure 2.28

where q_i^c is the eigenvector associated with the point C. To this end, note first that (2.3.33) does, indeed, define a plane in the Λ^k-space, and evidently Λ_c^k ($k = 1, 2, \ldots, M$), the critical values of the loading parameters at C, satisfy (2.3.33) by virtue of (2.3.32), C being a point of equilibrium. To prove that (2.3.33) is tangent to the stability boundary at C, suppose the latter is given in the form $\overset{*}{\Lambda}{}^1 = \overset{*}{\Lambda}{}^1(\overset{*}{\Lambda}{}^m)$, say, in the vicinity of C where $m = 2, 3, \ldots, M$. This assumed solution should, then, satisfy the equilibrium equations (2.3.32) which can be written as

$$[U_{ij} - \overset{*}{\Lambda}{}^1(\overset{*}{\Lambda}{}^m)E_{ij}^1 - \overset{*}{\Lambda}{}^m E_{ij}^m]q_i q_j = 0 \qquad (2.3.34)$$

upon multiplying the *i*th equation in (2.3.32) by q_i and adding the N equations. Since (2.3.34) is stationary with respect to changes in the q_i in the vicinity of an eigenvalue, partial differentiation of (2.3.34) with respect to any Λ^m yields upon evaluation at C

$$(-\overset{*}{\Lambda}{}^{1,m} E_{ij}^1 - E_{ij}^m) q_i^c q_j^c = 0 \qquad (2.3.35)$$

giving the slopes of the stability boundary as

$$\overset{*}{\Lambda}{}^{1,m}\Big|_C = -\frac{E_{ij}^m q_i^c q_j^c}{E_{ij}^1 q_i^c q_j^c} \qquad (2.3.36)$$

where $m = 2, 3, \ldots, M$. One immediately observes that the plane (2.3.33) has the same slopes and it is, therefore, tangential to the stability boundary at C.

2.3 Special critical points: asymmetric

Next, consider a ray

$$\Lambda^i = l^i \xi \tag{2.3.37}$$

emerging from the origin of the loading space and intersecting the tangent plane (2.3.33) as well as the critical surfaces (Figure 2.28). Here l^i are the direction cosines and ξ measures the distance from the origin. The intersection of this ray and the tangent plane yields the distance

$$\xi_t = \frac{U_{ij} q_i^c q_j^c}{E_{ij} q_i^c q_j^c} \quad \text{or} \quad \frac{1}{\xi_t} = \frac{E_{ij} q_i^c q_j^c}{U_{ij} q_i^c q_j^c} \tag{2.3.38}$$

where $E_{ij} = l^k E_{ij}^k$. Similarly, the intersection of the ray with the stability boundary yields

$$\xi_1 = \frac{U_{ij} q_i q_j}{E_{ij} q_i q_j} \quad \text{or} \quad \frac{1}{\xi_1} = \frac{E_{ij} q_i q_j}{U_{ij} q_i q_j} \tag{2.3.39}$$

where q_i is the eigenvector corresponding to ξ_1. The well-known extremum properties of the *Rayleigh Quotient*, then, result in

$$\frac{1}{\xi_1} \geq \frac{1}{\xi_t} \quad \text{or} \quad \xi_1 \leq \xi_t$$

since $\dfrac{1}{\xi_1}$ is the greatest eigenvalue associated with U_{ij} and E_{ij} [87 to 89].

Recalling that the point C was chosen arbitrarily on the stability boundary, one has the following theorem:

THEOREM 2.9. *The stability boundary cannot have convexity towards the region of stability.*

It seems that this theorem was first proved by Papkovich much earlier than the publication cited in Ref. [80] indicates. His proof is different from that presented here and a variation of it can be found in Ref. [83] where vibrating systems are considered (see also Section 2.4.3).

Evidently, the segment of a straight line joining any two points on the boundary represents a *lower bound* while a possible *upper bound* consists of segments of the tangent planes at the intercept of each coordinate axis [90].

2.4

Symmetric systems

Throughout this chapter symmetric systems will be treated in analogy with Chapter 1.3

2.4.1 Estimation of the stability boundary

Consider a multiple-parameter system described by the well-behaved total potential energy function

$$V = V(Q_i, z_\alpha, \Lambda^j) \tag{2.4.1}$$

which is symmetric in the z_α such that

$$V(Q_i, z_\alpha, \Lambda^j) = V(Q_i, -z_\alpha, \Lambda^j) \tag{2.4.2}$$

where z_α change sign as a set, and $i = 1, 2, \ldots, N$; $\alpha = 1, 2, \ldots, K$; $j = 1, 2, \ldots, M$.

In analogy with Section 1.3, it is assumed that the potential energy function (2.4.1) takes the form

$$\begin{aligned} V = &\frac{1}{2!}(U_{ij}Q_iQ_j + U_{\alpha\beta}z_\alpha z_\beta) \\ &+ \frac{1}{3!}(U_{ijk}Q_iQ_jQ_k + 3U_{i\alpha\beta}Q_iz_\alpha z_\beta) \\ &+ \frac{1}{4!}(\cdots) + \cdots \\ &- \Lambda^j E_i^j Q_i \end{aligned} \tag{2.4.3}$$

when referred to the origin of load-deflection space.

Introducing the orthogonal transformations

$$Q_i = a_{ij}u_j \quad \text{and} \quad z_\alpha = b_{\alpha\beta}y_\beta$$

the function V can be transformed into

$$W(u_i, y_\alpha, \Lambda^k) \equiv V(a_{ij}u_j, b_{\alpha\beta}y_\beta, \Lambda^k) \tag{2.4.4}$$

2.4 Symmetric systems

with the essential properties

$$W_{ij}=0 \quad \text{for} \quad i \neq j \quad \text{and} \quad W_{\alpha\beta}=0 \quad \text{for} \quad \alpha \neq \beta, \tag{2.4.5}$$

$$W_\alpha = W_{\alpha\beta\gamma} = \cdots = W_{i\alpha} = \cdots W_\alpha^k = \cdots = 0 \tag{2.4.6}$$

and

$$W_{ij}^k = W_{ijk}^l = \cdots = W_i^{jk} = W_{ij}^{kl} = \cdots = 0 \tag{2.4.7}$$

by virtue of the formulation. It is further noted that *any* combined differentiation with respect to z_α and Λ^i yields a zero derivative.

The stability boundary can be estimated in various ways. Firstly, a one-parameter perturbation technique very much like the one described in Chapter 1.3 can be adopted to estimate the boundary *point by point* [50]. To this end, one simply considers a ray

$$\Lambda^i = l^i \xi$$

directed from the origin, and treats the variable parameter ξ as the loading parameter Λ of Sections 1.3.3 and 1.3.4. All the formulae [e.g., equation (1.3.15)] derived in the equilibrium analysis in that section, then, become readily available provided it is recognized that the coefficient W_1^i of Section 1.3.3 is now given as

$$W_1' = W_1^i l^i.$$

Since this particular coefficient does not appear in the formulae associated with the stability analysis, such formulae can be used as given in Section 1.3.4 directly [e.g., equation (1.3.29)]. Furthermore, it is observed that for a different ray, the entire analysis need not be repeated and the change in the direction cosines can easily be accounted for by inserting the corresponding coefficient W_1^i in the related formulae. Thus, the flexibility and adaptability of the explicit formulae-type results derived in Chapter 1.3 for symmetric systems is noted once more in passing. Several significant specific problems can this way readily be treated [50] (see Section 2.5.2).

Secondly, the stability boundary can be estimated *as a whole* rather than *point by point* via a multiple-parameter perturbation technique. This can be achieved either by following the general procedure outlined in Section 2.2.1 or by developing a similar procedure for symmetric systems specifically. It was seen in Section 1.3.4 that the stability determinant evaluated on the fundamental path uncouples, allowing for examining each of the two components separately and more conveniently, and it is this property of symmetric systems that is worth exploiting.

Thus, suppose the equilibrium equations $W_i = W_\alpha = 0$ are solved

2.4.1 Estimation of the stability boundary

simultaneously to yield the fundamental equilibrium surface in the parametric form

$$u_i = u_i(\eta^k)$$
$$\Lambda^j = \Lambda^j(\eta^k) \qquad (2.4.8)$$

where $k = 1, 2, \ldots, M$.

It is understood that the fundamental surface lies in the $u_i - \Lambda^j$ subspace and deflections represented by y_α can only develop at special critical points where intersection of two surfaces is involved.

Substituting the assumed solution (2.4.8) into the equilibrium equations $W_i = 0$ yields the identities

$$W_i[u_j(\eta^l), \Lambda^k(\eta^l), 0] \equiv 0 \qquad (2.4.9)$$

Differentiating and evaluating at the origin successively results in

$$W_{ii} u_i^k + W_i^j \Lambda^{j,k} = 0,$$
$$W_{ijk} u_j^m u_k^n + W_{ii} u_i^{mn} + W_i^j \Lambda^{j,mn} = 0, \qquad (2.4.10)$$

etc.

Selecting appropriate basic variables as η^k, these ordered equations can readily be solved for the surface derivatives which are, then, used to construct the ordered equations of the fundamental surface in the form (2.2.8).

Consider now the stability determinant evaluated on the fundamental surface ($y_\alpha = 0$); since $\partial^2 W / \partial u_i \partial y_\alpha |_{y_\alpha = 0} = 0$, the stability determinant takes the form

$$\Delta(u_i) \equiv \Delta^u(u_i) \cdot \Delta^y(u_i) \equiv \left| \frac{\partial^2 W}{\partial u_i \partial u_j} \right| \cdot \left| \frac{\partial^2 W}{\partial y_\alpha \partial y_\beta} \right| \qquad (2.4.11)$$

where both determinants are functions of u_i only.

Assuming that the loss of stability is associated with simple critical points, it is seen that either Δ^u or Δ^y vanishes. The determinant

$$\Delta^u(u_k) \equiv \det |W_{ij}(u_k)| \qquad (2.4.12)$$

vanishes at *general* critical points, and the determinant

$$\Delta^y(u_k) \equiv \det |W_{\alpha\beta}(u_k)| \qquad (2.4.13)$$

at *special* critical points.

Suppose one is interested in the stability boundary associated with special critical points; assuming then that the equilibrium surface is obtained in the form

$$u_s = u_s(u_1, \Lambda^m) \qquad (m = 2, 3, \ldots, M)$$
$$\Lambda^1 = \Lambda^1(u_1, \Lambda^m) \qquad (2.4.14)$$

173

2.4 Symmetric systems

from (2.4.10), this solution is substituted into the determinant (2.4.13) to get

$$A^y(u_1, \Lambda^m) \equiv \Delta^y[u_i(u_1, \Lambda^m)] \equiv \det|W_{\alpha\beta}[u_i(u_1, \Lambda^m)]| \quad (2.4.15)$$

Differentiating (2.4.15) once with respect to u_1 and once with respect to Λ^m ($m \neq 1$) and evaluating at the origin yields

$$A_1^y = \Delta_i^y u_{i,1}$$
$$(A^y)^m = \Delta_i^y u_i^m$$

respectively. Here $u_{i,1}$ and u_i^m are the surface derivatives of the equilibrium analysis.

A second perturbation results in

$$A_{11}^y = \Delta_{ij}^y u_{i,1} u_{j,1} + \Delta_i^y u_{i,11}$$
$$(A^y)^{mn} = \Delta_{ij}^y u_i^m u_j^n + \Delta_i^y u_i^{mn} \quad (2.4.16)$$
$$(A^y)_1^m = \Delta_{ij}^y u_{i,1} u_j^m + \Delta_i^y u_{i,1}^m$$

These derivatives are, then, used to construct the ordered stability equations

$$A_0^y + A_1^y u_1 + (A^y)^m \Lambda^m = 0,$$
$$A_0^y + A_1^y u_1 + (A^y)^m \Lambda^m + \frac{1}{2!}[A_{11}^y u_1^2 + 2(A^y)_1^m u_1 \Lambda^m \quad (2.4.17)$$
$$+ (A^y)^{mn} \Lambda^m \Lambda^n] = 0,$$

etc.

where $A_0^y = \prod_{\alpha=1}^{K} W_{\alpha\alpha}$. Solving (2.4.17) with (2.4.14) concurrently, then, yields ordered estimates of the stability boundary associated with special points.

Another set of stability equations similar to (2.4.17) is obtained by evaluating the determinant (2.4.12) on the equilibrium surface (2.4.14), giving the general critical points. The procedure is now clear, and the derivation of these equations is left to the reader.

Finally, it must be noted that in analogy with Chapter 1.3, it is consistent to solve a second order stability equation, for example, with a third order equilibrium equation concurrently, this being due to the fact that only one part of the uncoupled stability determinant was considered in the derivation of the stability equations. It will be seen in the example of an arch that, if this property of symmetric systems is not utilized, then, corresponding to a third order equilibrium equation one must consider the fourth order stability equation (Section 2.5.3).

2.4.2 Stability boundary of a system with limited degrees of freedom

Problem. Assuming that the role of η^k has been given to u_1 and Λ^m ($m = 2, 3, \ldots, M$), solve the ordered equations (2.4.10) for surface derivatives to obtain explicit expressions and construct the third order equations of the equilibrium surface.

2.4.2 Stability boundary of a system with limited degrees of freedom

Consider the symmetric system described by the total potential energy function (2.4.4), and introduce a restriction on the number of degrees of freedom by assuming that $N = 1$. The fundamental equilibrium surface is, then, obtained easily as

$$y_\alpha = 0 \quad (\alpha = 1, 2, \ldots, K)$$
$$W_{11}u_1 + \frac{1}{2!} W_{111}u_1^2 + \frac{1}{3!} W_{1111}u_1^3 + \cdots = \Lambda^j W_1^j \tag{2.4.18}$$

from (2.4.10) or directly from the expansion of the function W.

The stability determinant of the system evaluated for $y_\alpha = 0$ takes the form

$$\Delta = \frac{\partial^2 W}{\partial u_1^2} \cdot \det \left| \frac{\partial^2 W}{\partial y_\alpha \partial y_\beta} \right| \tag{2.4.19}$$

Setting

$$\Delta^u = \frac{\partial^2 W}{\partial u_1^2} = 0, \tag{2.4.20}$$

one obtains the higher order equation in u_1

$$W_{11} + W_{111}u_1 + \frac{1}{2!} W_{1111}u_1^2 + \cdots = 0, \tag{2.4.21}$$

and introducing the roots of this equation into (2.4.18) yields the critical surfaces including the stability boundary. Obviously, all these surfaces are in the form of some multi-dimensional planes, and one has the following theorem for the systems under consideration:

THEOREM 2.10. *In the composition of the stability boundary associated with general critical points there can be no curved surfaces.*

Suppose now that

$$\Delta^y \equiv \det \left| \frac{\partial^2 W}{\partial y_\alpha \partial y_\beta} \right| \equiv \det |W_{\alpha\beta}(u_1)|$$
$$\equiv \det |W_{\alpha\beta} + \tfrac{1}{2} W_{1\alpha\beta}u_1 + \cdots| = 0 \tag{2.4.22}$$

175

2.4 Symmetric systems

It is seen that (2.4.22) yields another higher order equation in u_1, and introducing the roots of this equation into (2.4.18) results in the critical surfaces associated with special critical points which are again in the form of some multi-dimensional planes. Hence, the following theorem can be stated:

THEOREM 2.11. *In the composition of the stability boundary associated with special critical points there can be no curved surfaces.*

2.4.3 Convexity of the stability boundary of systems associated with a fundamental equilibrium plane

Instead of introducing a restriction on the degrees of freedom, consider now the truncated form of the potential energy function

$$V = \frac{1}{2!}(U_{ij}Q_iQ_j + U_{\alpha\beta}z_\alpha z_\beta)$$

$$+ \frac{3}{3!}U_{i\alpha\beta}Q_iz_\alpha z_\beta + \frac{1}{4!}U_{\alpha\beta\gamma\delta}z_\alpha z_\beta z_\gamma z_\delta - \Lambda^j E_i^j Q_i \quad (2.4.23)$$

In this form, the potential energy function (2.4.23) yields a *plane* fundamental equilibrium surface rather than a curved one, and the systems exhibiting nonlinear curved surfaces, like shallow arches, are out of the scope of this formulation. Various significant buckling problems concerning frames and plates under combined loading, however, are in this class.

The $N+K$ equilibrium equations $\partial W/\partial Q_i = 0$ and $\partial W/z_\alpha = 0$ yield

$$U_{ij}Q_j + \tfrac{1}{2}U_{i\alpha\beta}z_\alpha z_\beta - \Lambda^j E_i^j = 0$$

and $\quad(2.4.24)$

$$U_{\alpha\beta}z_\beta + U_{i\alpha\beta}Q_iz_\beta + \frac{1}{3!}U_{\alpha\beta\gamma\delta}z_\beta z_\gamma z_\delta = 0$$

respectively, which can be solved simultaneously to give the fundamental surface

$$\left.\begin{array}{l}z_\alpha = 0 \\ U_{ij}Q_j - \Lambda^j E_i^j = 0\end{array}\right\} \quad (2.4.25)$$

which is obviously in the form of a plane in the $Q_i - \Lambda^i$ subspace.

The critical points on the fundamental plane, which are now necessarily *special* points, can be located by examining the stability determinant

$$\Delta^z \equiv \det\left|\frac{\partial^2 V}{\partial z_\alpha \partial z_\beta}\right| \equiv \det|U_{\alpha\beta} + U_{i\alpha\beta}Q_i| \quad (2.4.26)$$

2.4.3 Convexity of the stability boundary of systems

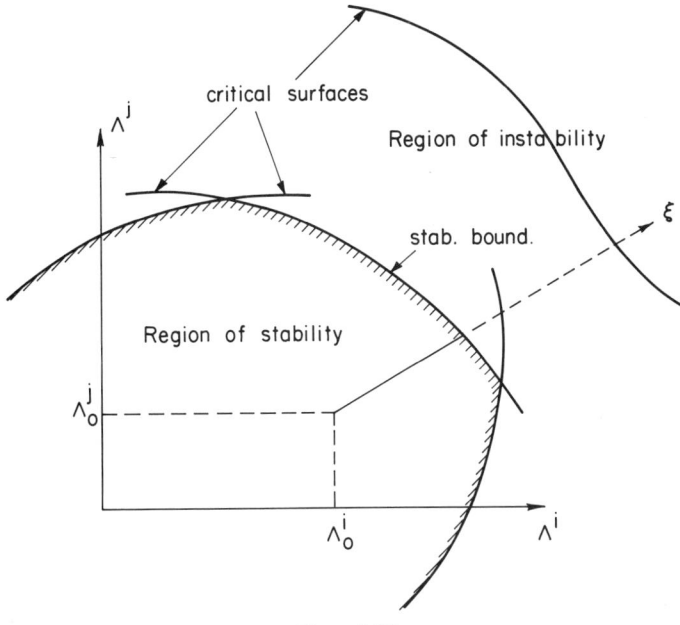

Figure 2.29

along the plane. It is noted that the determinant

$$\Lambda'' \equiv \det \left| \frac{\partial^2 V}{\partial Q_i \partial Q_j} \right| \equiv \det |U_{ij}|, \quad (2.4.27)$$

belonging to a distinct and uncoupled part of the quadratic form of the strain energy function, is a positive constant.

Substituting for Q_i from (2.4.25) into (2.4.26) and equating to zero, then, yields a determinantal equation in Λ^i only which, in general, defines some $(M-1)$ dimensional curved surfaces (the critical surfaces) in the load-space. It will be proved that the stability boundary is always concave with respect to the origin or the region of stability (Figure 2.29).

For this purpose, consider an arbitrarily chosen point $\Lambda^i = \Lambda_0^i$ ($i = 1, 2, \ldots, M$) in the *region of stability*. Corresponding to this point there exists a stable state of equilibrium (Q_i^0, Λ_0^i) on the fundamental plane in the load-deflection space (Figure 2.30).

Consider next the ray

$$\Lambda^i = \Lambda_0^i + l^i \xi \quad (2.4.28)$$

emanating from the chosen point in an arbitrary direction defined by the direction cosines l^i; ξ is the radius vector and will be allowed to take positive values only (Figure 2.29). The total potential energy of the system

2.4 Symmetric systems

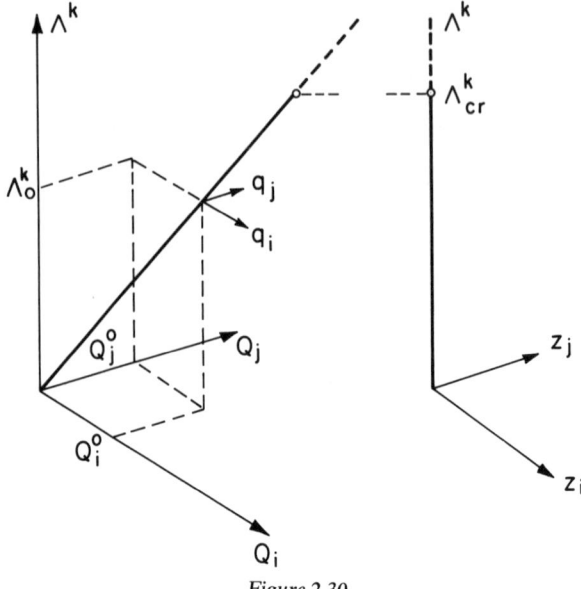

Figure 2.30

will be referred to this ray by introducing the transformation of translation

$$Q_i = Q_i^0 + q_i. \tag{2.4.29}$$

To facilitate the following analysis, one can also introduce the orthogonal transformation

$$q_i = a_{ij} u_j \tag{2.4.30}$$

such that the quadratic form of the energy function in q_i is diagonalized.

Thus, introducing (2.4.28) to (2.4.30) in the energy function (2.4.23), remembering that the point (Q_i^0, Λ_0^j) is one of equilibrium and ignoring irrelevant terms one gets

$$V = \tfrac{1}{2}(V_{ii} u_i^2 + V_{\alpha\beta} z_\alpha z_\beta)$$
$$+ \tfrac{1}{2} V_{i\alpha\beta} u_i z_\alpha z_\beta - V_i' u_i \xi \tag{2.4.31}$$

along the ray.

The fourth order terms in z_α are dropped because they do not influence the buckling behaviour although they play an important role in the post-buckling range.

The equilibrium equations $\partial V/\partial u_i = 0$ and $\partial V/\partial z_\alpha = 0$ now yield the fundamental path

$$z_\alpha = 0$$
$$u_i = \frac{V_i'}{V_{ii}} \xi \tag{2.4.32}$$

2.4.3 Convexity of the stability boundary of systems

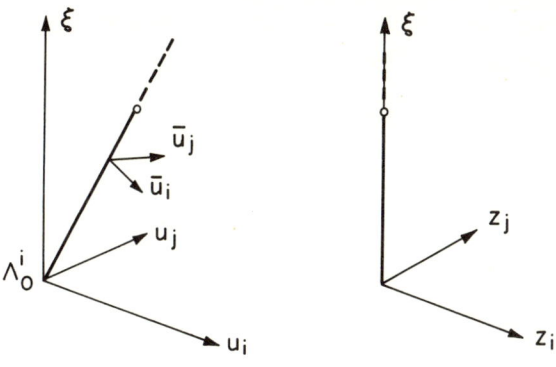

Figure 2.31

along the ray (2.4.28). In Figure 2.31 this situation is illustrated for the case in which the ray intersects the stability boundary. It is noted, however, that depending on the direction of the ray, the stability boundary might not be intersected at all, the equilibrium path maintaining its stability along the ray.

At this stage, it is required to introduce the sliding coordinates $\bar{u}_i$ such that

$$u_i = \frac{V'_i}{V_{ii}} \xi + \bar{u}_i \tag{2.4.33}$$

Substituting (2.4.33) into (2.4.31) and ignoring irrelevant terms one has

$$V = \tfrac{1}{2}(V_{ii}\bar{u}_i^2 + V_{\alpha\beta}z_\alpha z_\beta)$$
$$+ \tfrac{1}{2}(V_{i\alpha\beta}\bar{u}_i z_\alpha z_\beta + V'_{\alpha\beta}\xi z_\alpha z_\beta) \tag{2.4.34}$$

where $V'_{\alpha\beta} = V_{i\alpha\beta} \dfrac{V'_i}{V_{ii}}$.

Since the point (Q_i^0, Λ_0^j) was chosen in the region of stability the quadratic form

$$V_{\alpha\beta}z_\alpha z_\beta$$

is positive definite and hence there exists a linear transformation

$$z_\alpha = b_{\alpha\beta} y_\beta, \quad \det|b_{\alpha\beta}| \neq 0$$

which simultaneously diagonalizes the two quadratic forms

$$V_{\alpha\beta}z_\alpha z_\beta \quad \text{and} \quad V'_{\alpha\beta}\xi z_\alpha z_\beta.$$

2.4 Symmetric systems

Substituting for z_α into (2.4.34) one finally gets

$$V = \tfrac{1}{2}(V_{ii}\bar{u}_i^2 + \bar{V}_{\alpha\alpha} y_\alpha^2)$$
$$+ \tfrac{1}{2}(\bar{V}_{i\alpha\beta}\bar{u}_i y_\alpha y_\beta + \bar{V}'_{\alpha\alpha}\xi y_\alpha^2) \qquad (2.4.35)$$

The equilibrium equations $V_i = V_\alpha = 0$ now yield the fundamental solution

$$\bar{u}_i = 0$$
$$y_\alpha = 0 \qquad (2.4.36)$$

as expected, and the stability determinant takes the form

$$\Delta = \det|V_{ii}| \cdot \det|\bar{V}_{\alpha\alpha} + \xi \bar{V}'_{\alpha\alpha}| \qquad (2.4.37)$$

where all coefficients V_{ii} and $\bar{V}_{\alpha\alpha}$ are positive while $\bar{V}'_{\alpha\alpha}$ can be positive or negative depending on the direction of the ray.

Suppose, corresponding to a certain ray, at least one of $V'_{\alpha\alpha}$, $V'_{\gamma\gamma}$ say, is negative, then proceeding along the chosen ray, i.e., increasing the positive quantity ξ from zero, one is bound to reach a point at which the stability coefficient $\bar{V}_{\gamma\gamma} + \xi V'_{\gamma\gamma}$ vanishes resulting in a vanishing stability determinant and a critical parameter

$$\xi_c = -\frac{V_{\gamma\gamma}}{V'_{\gamma\gamma}} \qquad (2.4.38)$$

associated with the chosen ray. If there are some other negative coefficients $\bar{V}'_{\alpha\alpha}$, then, as one proceeds along the ray, the corresponding stability coefficients vanish upon intersecting the associated critical surfaces. If (2.4.38) is the smallest critical value of ξ it corresponds to an intersection with the stability boundary and the entire ray beyond this point lies in the unstable region.

On the other hand, if all $\bar{V}'_{\alpha\alpha}$ are positive, then, all the stability coefficients remain positive as the quantity ξ is increased from zero, and consequently, a loss of stability along the corresponding ray is ruled out, the stability boundary not being intersected.

It is clear that a ray in an arbitrary direction from an arbitrary point in the region of stability cannot intersect the stability boundary associated with a principal coordinate y_α more than once, and if the stability of the system is once lost it cannot be regained by proceeding along the ray. In other words, once the region of stability is left, by proceeding along the same ray, it cannot be re-entered. The obvious result of these arguments is as follows:

THEOREM 2.12. *The stability boundary cannot have convexity towards the region of stability.*

2.4.4 Equilibrium surface in the vicinity of a symmetric critical point

It is worth noting that this theorem cannot be proved for a more general class of systems associated with nonlinear (curved) fundamental surfaces. This is because in the formulation of such a general problem the involvement of the third and higher order terms in Q_i results in a function $V(\bar{u}_i, y_\alpha, \xi)$ which is nonlinear in ξ. If the fundamental surface of a particular system is not highly nonlinear, however, it can be expected that Theorem 2.12 is obeyed.

2.4.4 Equilibrium surface in the vicinity of a symmetric special critical point

In this section, general post-critical characteristics of the symmetric systems described by the total potential energy function (2.4.1) will be studied.

It is known that the $N+K$ equilibrium equations $V_i = V_\alpha = 0$ can be solved simultaneously to yield the fundamental surface

$$Q_i = Q_i^F(\Lambda^j), \quad z_\alpha^F = 0 \qquad (2.4.39)$$

It will now be assumed that in the region of interest, this surface is single-valued so that the correspondence between a set of Λ^i and Q_i^F is unique. In analogy with Section 2.3.1, the potential energy can, then, be referred to the fundamental surface (2.4.39) by setting

$$Q_i = Q_i^F(\Lambda^j) + q_i \qquad (2.4.40)$$

Two further changes of coordinates by means of the linear orthogonal transformations

$$\begin{aligned} q_i &= a_{ij}(\Lambda^k) u_j, & a_{ij}(\Lambda^l) a_{jk}(\Lambda^l) &= \delta_{ik} \\ z_\alpha &= b_{\alpha\beta}(\Lambda^k) y_\beta, & b_{\alpha\beta}(\Lambda^l) b_{\beta\gamma}(\Lambda^l) &= \delta_{\alpha\gamma} \end{aligned} \qquad (2.4.41)$$

are introduced to diagonalize the quadratic forms of the energy function in q_i and z_α respectively. Introducing (2.4.40) and (2.4.41) into the energy function $V(Q_i, z_\alpha, \Lambda^j)$ one gets the new function

$$S(u_i, y_\alpha, \Lambda^j) \equiv V[Q_i^F(\Lambda^k) + a_{ij}(\Lambda^k) u_j, 0 + b_{\alpha\beta} y_\beta, \Lambda^k] \qquad (2.4.42)$$

with the properties

$$\left. \begin{aligned} S_i(0, 0, \Lambda^k) &= S_i^j(0, 0, \Lambda^k) = S_i^{jk}(0, 0, \Lambda^l) = \cdots = 0 \\ S_\alpha(0, 0, \Lambda^k) &= S_\alpha^i(0, 0, \Lambda^k) = S_\alpha^{ij}(0, 0, \Lambda^k) = \cdots = 0 \end{aligned} \right\} \qquad (2.4.43)$$

and

$$\left. \begin{aligned} S_{ij}(0, 0, \Lambda^k) &= S_{ij}^k(0, 0, \Lambda^l) = \cdots = 0 & \text{for} \quad i \neq j \\ S_{\alpha\beta}(0, 0, \Lambda^k) &= S_{\alpha\beta}^i(0, 0, \Lambda^k) = \cdots = 0 & \text{for} \quad \alpha \neq \beta \end{aligned} \right\} \qquad (2.4.44)$$

2.4 Symmetric systems

which follow from the fact that the $u_i = y_\alpha = 0$ define the fundamental surface, and the quadratic forms of energy are diagonalized at every point of this surface in the region of interest.

It can further be shown that the symmetry property (2.4.2) results in

$$S_\alpha = S_{\alpha\beta\gamma} = \cdots = S_{\alpha i} = \cdots = S_\alpha^i = \cdots = 0 \qquad (2.4.45)$$

Considering the $N+K$ stability coefficients $S_{ii}(0, 0, \Lambda^j)$ and $S_{\alpha\alpha}(0, 0, \Lambda^j)$, attention will be focussed on a discrete critical point F at which one of the latter coefficients, S_{11} say, vanishes. Finally, introduce the orthogonal transformation of the Λ^i coordinates

$$\Lambda^i = \gamma^{ij} \Phi^j, \qquad \gamma^{ij}\gamma^{ik} = \delta^{ik} \qquad (2.4.46)$$

so that when (2.4.46) is substituted in the function (2.4.42), the resulting function

$$\Psi(u_i, y_\alpha, \Phi^j) \equiv S(u_i, y_\alpha, \gamma^{ij}\Phi^j) \qquad (2.4.47)$$

has the properties

$$\Psi_{11}^1(0, 0, \overset{*}{\Phi}{}^j_F) \neq 0, \qquad \Psi_{11}^m(0, 0, \overset{*}{\Phi}{}^j_F) = 0 \qquad (2.4.48)$$

where $m = 2, 3, \ldots, M$, the subscripts on Ψ denote partial differentiation with respect to the critical coordinate y_1, and the $\overset{*}{\Phi}{}^j_F$ are the critical values of Φ^j at F.

It can readily be shown that the properties (2.4.43) to (2.4.45) are now replaced by

$$\left. \begin{array}{l} \Psi_i(0, 0, \Phi^k) = \Psi_i^j(0, 0, \Phi^k) = \cdots = 0 \\ \Psi_\alpha(0, 0, \Phi^k) = \Psi_\alpha^j(0, 0, \Phi^k) = \cdots = 0 \end{array} \right\} \qquad (2.4.49)$$

$$\left. \begin{array}{ll} \Psi_{ij}(0, 0, \Phi^k) = \Psi_{ij}^k(0, 0, \Phi^l) = \cdots = 0 & \text{for } i \neq j \\ \Psi_{\alpha\beta}(0, 0, \Phi^k) = \Psi_{\alpha\beta}^k(0, 0, \Phi^l) = \cdots = 0 & \text{for } \alpha \neq \beta \end{array} \right\} \qquad (2.4.50)$$

and

$$\Psi_\alpha = \Psi_{\alpha\beta\gamma} = \cdots \Psi_{\alpha i} = \cdots \Psi_\alpha^i = \cdots = 0 \qquad (2.4.51)$$

respectively.

The function $\Psi(u_i, y_\alpha, \Phi^j)$, the only necessary properties of which are given by (2.4.48) to (2.5.51) will now be used to explore the neighbourhood of the critical point F at which

$$\Psi_{11}(0, 0, \overset{*}{\Phi}{}^j_F) = 0, \qquad \Psi_{\delta\delta}(0, 0, \overset{*}{\Phi}{}^j_F) \neq 0 \quad \text{for} \quad \delta = 2, 3, \ldots, K$$

and $\qquad (2.4.52)$

$$\Psi_{ii}(0, 0, \overset{*}{\Phi}{}^j) \neq 0$$

2.4.4 Equilibrium surface in the vicinity of a symmetric critical point

The transformations introduced enable us to choose the independent parameters as y_1 and Φ^m ($m \neq 1$) at the onset of the analysis, and assume the post-buckling surface in the vicinity of F in the form

$$u_i = u_i(y_1, \Phi^m), \qquad y_\delta = y_\delta(y_1, \Phi^m)$$

and (2.4.53)

$$\Phi^1 = \Phi^1(y_1, \Phi^m)$$

Substituting these functions back into the equilibrium equations $\Psi_i = \Psi_\alpha = 0$, one has the identities

$$\Psi_i[u_j(y_1, \Phi^m), y_\delta(y_1, \Phi^m), \Phi^1(y_1, \Phi^m), y_1, \Phi^m] \equiv 0 \qquad (2.4.54)$$

$$\Psi_\alpha[u_i(y_1, \Phi^m), y_\delta(y_1, \Phi^m), \Phi^1(y_1, \Phi^m), y_1, \Phi^m] \equiv 0 \qquad (2.4.55)$$

Differentiating (2.4.54) once with respect to u_1 and once with respect to Φ^m ($m = 2, 3, \ldots, M$) yields

$$\Psi_{ij} u_{j,1} + \Psi_{i\delta} y_{\delta,1} + \Psi_i^1 \Phi_1^1 + \Psi_{i1} = 0$$

and (2.4.56)

$$\Psi_{ij} u_j^m + \Psi_{i\delta} y_\delta^m + \Psi_i^1 \Phi^{1,m} + \Psi_i^m = 0$$

Evaluating these equations at F results in

$$u_{i,1} = 0 \quad \text{and} \quad u_i^m = 0. \qquad (2.4.57)$$

Differentiating (2.4.55) once with respect to y_1 and once with respect to Φ^m one has

$$\Psi_{\alpha i} u_{i,1} + \Psi_{\alpha \delta} y_{\delta,1} + \Psi_\alpha^1 \Phi_1^1 + \Psi_{\alpha 1} = 0$$

and

$$\Psi_{\alpha i} u_i^m + \Psi_{\alpha \delta} y_\delta^m + \Psi_\alpha^1 \Phi^{1,m} + \Psi_\alpha^m = 0$$

which upon evaluation at F one obtains

$$y_{\delta,1} = 0, \qquad y_\delta^m = 0 \quad (\delta \neq 1) \qquad (2.4.58)$$

Differentiating (2.4.54) with respect to y_1 for a second time yields

$$(\cdots) u_{j,1} + \Psi_{ij} u_{j,11} + (\cdots) y_{\delta,1} + \Psi_{i\delta} y_{\delta,11}$$
$$+ (\Psi_{ij}^1 u_{j,1} + \Psi_{i\delta}^1 y_{\delta,1} + \Psi_i^{11} \Phi_1^1 + \Psi_{i1}^1) \Phi_1^1$$
$$+ \Psi_i^1 \Phi_{11}^1 + \Psi_{i1j} u_{j,1} + \Psi_{i1\delta} y_{\delta,1} + \Psi_{i1}^1 \Phi_1^1 + \Psi_{i11} = 0$$

which gives upon evaluation

$$u_{i,11} = -\Psi_{i11}/\Psi_{ii} \qquad (2.4.59)$$

Differentiation of (2.4.55) with respect to y_1 for a second time results

2.4 Symmetric systems

in, upon evaluation

$$\Phi_1^1 = 0 \quad \text{and} \quad y_{\delta,11} = 0 \tag{2.4.60}$$

Differentiate now (2.4.54) first with respect to y_1 and for a second time with respect to Φ^m to get

$$(\cdots)u_{j,1} + \Psi_{ij}u_{j,1}^m + (\cdots)y_{\delta,1} + \Psi_{i\delta}y_{\delta,1}^m + (\cdots)\Phi_1^1$$
$$+ \Psi_i^1\Phi_1^{1,m} + \Psi_{i1j}u_j^m + \Psi_{i1\delta}y_\delta^m + \Psi_{i1}^1\Phi^{1,m}$$
$$+ \Psi_{i1}^m = 0$$

which yields

$$u_{i,1}^m = 0 \tag{2.4.61}$$

Similarly, differentiations of (2.4.54) and (2.4.55) with respect to certain independent variables twice result in

$$\Phi^{1,m} = 0, \quad y_{\delta,1}^m = 0$$
$$u_i^{mn} = 0, \quad y_\delta^{mn} = 0 \tag{2.4.62}$$

Proceeding in the same manner, the third perturbation of the equations (2.4.54) and (2.4.55) yield, upon evaluation, the following derivatives:

$$\Phi_{11}^1 = -\frac{1}{3\Psi_{11}^1}\left[\Psi_{1111} - 3\sum_{i=1}^{N}\frac{(\Psi_{i11})^2}{\Psi_{ii}}\right],$$

$$\Phi^{1,mn} = -\Psi_{11}^{mn}/\Psi_{11}^1,$$

$$y_{\delta,1}^{mn} = 0, \quad \Phi_1^{1,m} = 0 \tag{2.4.63}$$

$$y_{\delta,111} = -\frac{1}{\Psi_{\delta\delta}}\left[\Psi_{\delta 111} - 3\frac{\Psi_{i\delta 1}\Psi_{i11}}{\Psi_{ii}}\right]$$

$$y_{\delta,11}^m = 0, \quad y_\delta^{mnr} = 0$$

$$u_{i,1}^{mn} = 0, \quad u_i^{mnr} = 0.$$

The asymptotic equations of the post-buckling surface (2.4.53) can now be constructed. Thus, using the derivatives (2.4.57) to (2.4.63) and $\varphi^i = \Phi^i - \overset{*}{\Phi}{}_F^i$ one gets

$$\varphi^1 = -\frac{1}{6\Psi_{11}^1}\left[\Psi_{1111} - 3\sum_{i=1}^{N}\frac{(\Psi_{i11})^2}{\Psi_{ii}}\right]y_1^2 - \frac{\Psi_{11}^{mn}}{2\Psi_{11}^1}\varphi^m\varphi^n$$

$$\equiv ay_1^2 + b^{mn}\varphi^m\varphi^n \tag{2.4.64}$$

$$u_i = -\frac{1}{2}\frac{\Psi_{i11}}{\Psi_{ii}}y_1^2 \tag{2.4.65}$$

$$y_\delta = -\frac{1}{3!}\frac{1}{\Psi_{\delta\delta}}\left(\Psi_{\delta 111} - 3\frac{\Psi_{i\delta 1}\Psi_{i11}}{\Psi_{ii}}\right)y_1^3 \tag{2.4.66}$$

in the vicinity of F.

2.4.4 Equilibrium surface in the vicinity of a symmetric critical point

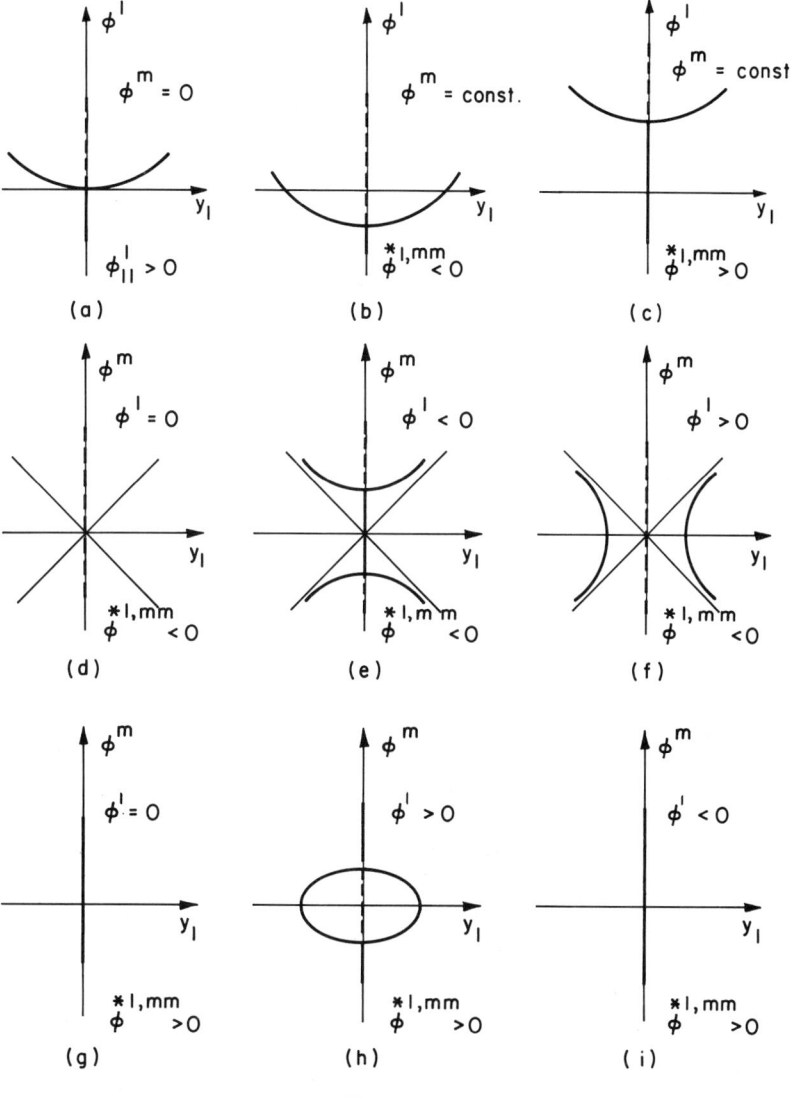

Figure 2.32

Various equilibrium paths lying on the fundamental and post-buckling surfaces, (i.e., various intersections of these surfaces) are depicted in Figures 2.32 and 2.33.

To facilitate understanding, consider a ray defined by

$$\varphi^i = l^i \xi \quad \text{where} \quad l^1 \neq 0;$$

2.4 Symmetric systems

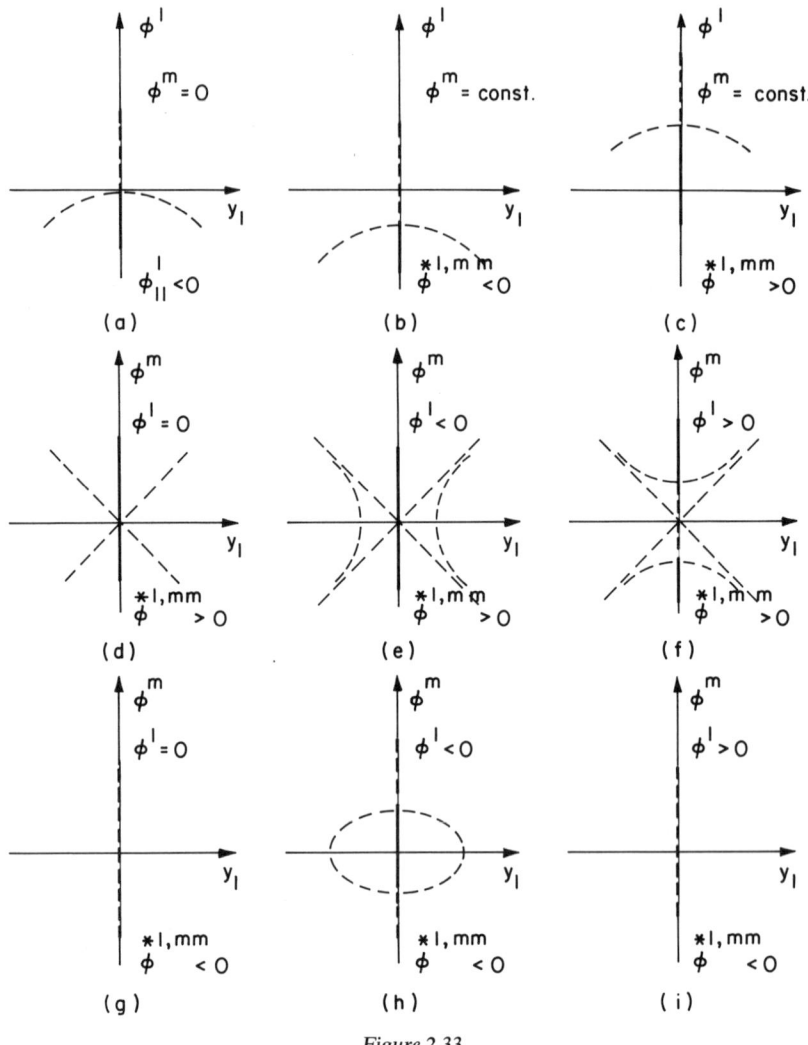

Figure 2.33

equation (2.4.64), then, takes the form

$$\xi = \frac{a}{l^1} y_1^2 \tag{2.4.67}$$

to a first order approximation. (2.4.67) indicates a *symmetric point of bifurcation* on a plot of ξ versus y_1. On the other hand, if one specifies $l^1 = 0$, (2.4.64) yields

$$a y_1^2 + b^{mn} l^m l^n \xi^2 = 0 \tag{2.4.68}$$

2.4.4 Equilibrium surface in the vicinity of a symmetric critical point

which defines either the point F or two intersecting lines depending on the signs of coefficients. The latter is, then, the phenomenon mentioned in Section 1.1.6, and it is illustrated in Figure 2.34 (see also Figures 2.32d and 2.33d). One notes that this type of behaviour is produced when a loading ray in the φ^m subspace is followed, and the post buckling surface is *anticlastic*. It will be seen in the next section that $\varphi^1 = 0$ defines the plane tangent to the stability boundary at F, and it follows that a loading path which is tangential to the stability boundary results in two intersecting post-buckling paths with finite slopes provided the curvatures of the surface in the $\varphi^1 - y_1$ and $\varphi^1 - \varphi^m$ (or $\varphi^1 - \xi$) planes have opposite signs. The equation of the post-buckling paths for a particular φ^m is obtained from (2.4.64) as

$$\left[\Psi_{1111} - 3 \sum_{i=1}^{N} \frac{(\Psi_{i11})^2}{\Psi_{ii}} \right] y_1^2 + 3\Psi_{11}^{mn}(\varphi^m)^2 = 0 \qquad (2.4.69)$$

Since $\varphi^1 = 0$ is equivalent to $\Psi_{11}^1 = 0$, in one-parameter *symmetric* systems, this phenomenon occurs when the derivative of the vanishing stability coefficient with respect to the loading parameter also vanishes simultaneously, and it may more aptly be termed *tri-furcation* rather than bifurcation. Such multi-furcations were seen to occur at coincident critical points in Section 1.5.3 and the phenomenon was termed compound branching; the distinctive feature of *tri-furcation* here is that it occurs at a simple critical point.

If the post-buckling surface is *synclastic*, then, the paths with *finite* slopes cannot be obtained and the behaviour of the system in the vicinity of F is as illustrated in Figures 2.32g, h, i and Figures 2.33g, h, i.

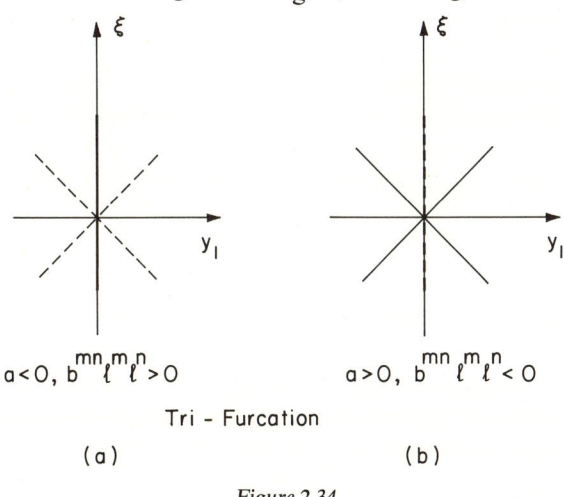

Tri - Furcation

(a)   (b)

Figure 2.34

187

2.4 Symmetric systems

It is thus observed that a symmetric system can exhibit post-buckling paths with finite slopes under some circumstances as well as with zero slopes, and one might wonder whether such a special critical point can be called *symmetric* in analogy with one-parameter systems. Since the paths are initially symmetric with respect to the loading as well as the y_1 axis, the special critical points here will be called *symmetric*. It is, then, understood that a symmetric special point generally appears as a symmetric point of *bifurcation* under proportional loading but under certain circumstances *tri-furcation* may be exhibited at the same point.

2.4.5 Stability boundary in the vicinity of a symmetric special critical point

It was proved in Section 2.3.2 that the stability boundary is, in fact, the intersection of the fundamental and initial post-buckling surfaces as Theorem 2.7 indicates. This is true for all special critical points, symmetric or asymmetric. It is not, therefore, required to perform a similar analysis here, since the intersection of the two surfaces can readily be obtained. Thus, setting $u_i = y_j = 0$ in the equations (2.4.64) to (2.4.66) of the post-buckling surface, one has the stability boundary

$$\overset{*}{\varphi}{}^1 = -\frac{\Psi_{11}^{mn}}{2\Psi_{11}^1} \overset{*}{\varphi}{}^m \varphi^n \tag{2.4.70}$$

in the vicinity of F provided F is primary so that

$$\Psi_{11} = 0, \quad \text{all} \quad \Psi_{\delta\delta} > 0 \; (\delta \neq 1), \quad \text{all} \quad \Psi_{ii} > 0$$

If F is not primary, then (2.4.70) describes a critical surface.

From the equations of the equilibrium surface and the stability boundary one cannot infer any definite information regarding the convexity of the stability boundary which may lead to a theorem. It seems that in this generality no such theorem can be stated. Nevertheless, one observes that if the matrix Ψ_{11}^{mn} is positive or negative definite, then, the stability boundary is *synclastic*, and convex towards the region of stability or instability respectively, provided $\Psi_{11}^1 < 0$.

2.4.6 Stability distribution on the equilibrium surfaces

Following the same procedure as in Section 2.3.3, the stability of the equilibrium states in the vicinity of F will now be examined.

To this end, introduce the stability determinant

$$\Delta(u_i, y_\alpha, \varphi^j) \equiv \det |\Psi_{uy}(u_i, y_\alpha, \varphi^j)| \tag{2.4.71}$$

2.4.6 Stability distribution on the equilibrium surfaces

where the subscripts u and y on Ψ denote differentiation with respect to the complete range of generalized coordinates u_i and y_α, i.e., $u, y = 1, 2, \ldots, N, 1, 2, \ldots, K$. It is understood that the stability determinant uncouples on the fundamental surface and takes the form $\Delta = \Delta^u \cdot \Delta^y$, however, for complete generality this is not being stressed here.

Differentiating (2.4.71) once with respect to u_i and once with respect to φ^1 and evaluating at F yields

$$\Delta_i = \Psi_{11i} \prod_{i=1}^{N} \Psi_{ii} \prod_{\delta=2}^{K} \Psi_{\delta\delta} \qquad (2.4.72)$$

and

$$\Delta^1 = \Psi_{11}^1 \prod_{i=1}^{N} \Psi_{ii} \prod_{\delta=2}^{K} \Psi_{\delta\delta} \qquad (2.4.73)$$

Differentiating (2.4.71) once with respect to y_α ($\alpha = 1, 2, \ldots, K$) and once with respect to φ^m ($m = 2, 3, \ldots, M$), one observes that $\Delta_\alpha = \Delta^m = 0$; a second differentiation yields on evaluation

$$\Delta^{mn} = \Psi_{11}^{mn} \prod_{i=1}^{N} \Psi_{ii} \prod_{\delta=2}^{K} \Psi_{\delta\delta},$$

$$\Delta_{\alpha\beta} = \left(\Psi_{11\alpha\beta} - 2 \frac{\Psi_{i1\alpha} \Psi_{i1\beta}}{\Psi_{ii}} \right) \prod_{i=1}^{N} \Psi_{ii} \prod_{2}^{K} \Psi_{\delta\delta} \qquad (2.4.74)$$

$$\Delta_\alpha^m = 0$$

Using these derivatives and Taylor's expansion one has

$$\Delta = \left[\Psi_{i11} u_i + \Psi_{11}^1 \varphi^1 + \frac{1}{2!} \Psi_{11}^{mn} \varphi^m \varphi^n \right.$$
$$\left. + \frac{1}{2!} \left(\Psi_{11\alpha\beta} - 2 \frac{\Psi_{i1\alpha} \Psi_{i1\beta}}{\Psi_{ii}} \right) y_\alpha y_\beta \right] \prod_{i=1}^{N} \Psi_{ii} \prod_{i=2}^{K} \Psi_{\delta\delta} + \cdots \qquad (2.4.75)$$

Evaluating (2.4.75) at an arbitrary state on the fundamental surface, i.e., setting $u_i = y_\alpha = 0$, yields to a first approximation

$$\Delta = (\Psi_{11}^1 \varphi^1 + \tfrac{1}{2} \Psi_{11}^{mn} \varphi^m \varphi^n) \prod_{i=1}^{N} \Psi_{ii} \prod_{\delta=2}^{K} \Psi_{\delta\delta}$$

which results in the following stability criterion:

$$\Psi_{11}^1 \varphi^1 + \tfrac{1}{2} \Psi_{11}^{mn} \varphi^m \varphi^n \begin{array}{l} > \quad \text{stable} \\ = 0 \text{ for critical equilibrium} \\ < \quad \text{unstable} \end{array} \qquad (2.4.76)$$

in the vicinity of F.

As in Section 2.3.3, consider a point A on the stability boundary;

2.4 Symmetric systems

then, for the points defined by

$$\varphi^1 = \overset{*}{\varphi}{}_A^1 + \tau$$
$$\varphi^m = \overset{*}{\varphi}{}_A^m, \qquad (2.4.77)$$

the stability criterion (2.4.76) takes the simple form

$$\begin{array}{ll} > & \text{stable} \\ \Psi_{11}^1 \tau = 0 & \text{for critical equilibrium} \qquad (2.4.78) \\ < & \text{unstable} \end{array}$$

where τ is again a small increment in φ^1.

If, for instance, $\Psi_{11}^1 < 0$, then $\tau < 0$ defines the *region of stability* and $\tau > 0$ the *region of instability* as A slides on the stability boundary. It is seen that the stability boundary divides the fundamental surface into stable and unstable domains.

In order to examine the stability of the states lying on the post-buckling surface, evaluate the determinant (2.4.75) on this surface which is defined by (2.4.64) to (2.4.66). Substituting for u_i, y_δ and φ^i into (2.4.75) yields to a first order approximation

$$\Delta = \frac{1}{3}\left[\Psi_{1111} - 3\sum_{i=1}^{N}\frac{(\Psi_{i11})^2}{\Psi_{ii}}\right] y_1^2 \prod_{i=1}^{N} \Psi_{ii} \prod_{\delta=2}^{K} \Psi_{\delta\delta} \qquad (2.4.79)$$

Evidently, the stability of the neighbouring equilibrium states does not depend on y_1. In fact, the sign of the expression in brackets in (2.4.79) determines the stability of the initial post-buckling surface *as a whole* so that if this expression is positive (negative) the surface in the vicinity of F is totally stable (unstable).

The stability determinant evaluated on the post-buckling surface can alternatively be expressed as

$$\Delta = -(2\Psi_{11}^1 \varphi^1 + \Psi_{11}^{mn}\varphi^m\varphi^n) \prod_{i=1}^{N}\Psi_{ii} \prod_{\delta=2}^{K} \Psi_{\delta\delta}$$

in which case the stability of the states defined by (2.4.77) is determined according to the following criterion:

$$\begin{array}{ll} > & \text{stable} \\ -\Psi_{11}^1 \tau = 0 & \text{for critical equilibrium} \qquad (2.4.80) \\ < & \text{unstable} \end{array}$$

Comparing (2.4.78) to (2.4.80), it may be concluded that to the points of the region of stability (instability) only unstable (stable) post-buckling equilibrium states can correspond. It is easy to show, however, that if the

2.4.6 Stability distribution on the equilibrium surfaces

post-buckling surface is stable (unstable), corresponding to the points of the region of stability (the region of instability) there exists no post-buckling equilibrium states. To this end, consider again the point A sliding on the stability boundary; the post-buckling states associated with (2.4.77) can be expressed in the form

$$y_1 = \pm \left(\frac{\tau}{a}\right)^{\frac{1}{2}} \qquad (2.4.81)$$

upon substitution for φ^1 and φ^m into the equilibrium equation (2.4.64). It, then, follows that if $a > 0$ ($a < 0$), real equilibrium states can only exist for $\tau > 0$ ($\tau < 0$). If, for instance, $\Psi_{11}^1 < 0$, then, by virtue of (2.4.79), $a > 0$ means a stable post-buckling surface in which case for $\tau < 0$ the post-buckling states are not real. In Figures 2.32 and 2.33 the stable paths are shown with full lines while broken lines indicate unstable paths as usual. It is observed in particular that the two post-buckling lines in Figure 2.34 are either stable as in (b) or unstable as in (a) in compliance with the above discussion. Thus, generally speaking, at a point of *tri-furcation* a stable path (fundamental) can continue to be stable after passing through a critical state where two unstable paths (post-buckling) are intersected. Similarly, an unstable path remains unstable upon passing through the point of *tri-furcation* where two stable paths are intersected.

On the basis of the foregoing theory, the following theorems concerning the stability distribution on the equilibrium surfaces in the vicinity of a symmetric special critical point are established:

THEOREM 2.13. *The initial post-buckling surface is either totally stable or unstable.*

THEOREM 2.14. *The stability boundary constitutes an existence boundary with regard to the post-buckling surface so that if this surface is stable (unstable), to the points of the region of stability (the region of instability) no post-buckling states can correspond.*

Finally, the stability of the symmetric special critical point F itself will be discussed. Since the second variation of energy is positive semi-definite, higher order variations of energy are required for a decision regarding stability, and such an analysis can be carried out by following the procedure outlined in Section 1.1.4.

Considering, then, an arbitrary path defined by

$$\begin{aligned} u_i &= u_i(\eta) \\ y_\alpha &= y_\alpha(\eta) \end{aligned} \qquad (2.4.82)$$

2.4 Symmetric systems

the variations of $\Psi(u_i, y_\alpha, \overset{*}{\Phi}{}^j_F)$ along this path can be examined [43, 48] by successively differentiating the change in energy

$$\psi(\eta) \equiv \Psi[u_i(\eta), y_\alpha(\eta), \overset{*}{\Phi}{}^j_F] - \Psi[0, 0, \overset{*}{\Phi}{}^j_F] \tag{2.4.83}$$

Thus

$$\frac{d\psi}{d\eta} = \Psi_i u_{i,\eta} + \Psi_\alpha y_{\alpha,\eta}$$

and

$$\left.\frac{d\psi}{d\eta}\right|_F = 0$$

Similarly,

$$\frac{d^2\psi}{d\eta^2} = (\Psi_{ij} u_{j,\eta} + \Psi_{i\alpha} y_{\alpha,\eta}) u_{i,\eta} + \Psi_i u_{i,\eta\eta}$$

$$+ (\Psi_{\alpha i} u_{i,\eta} + \Psi_{\alpha\beta} y_{\beta,\eta}) y_{\alpha,\eta} + \Psi_\alpha y_{\alpha,\eta\eta}$$

and

$$\left.\frac{d^2\psi}{d\eta^2}\right|_F = \Psi_{ii}(u_{i,\eta})^2 + \Psi_{\delta\delta}(y_{\delta,\eta})^2 \tag{2.4.84}$$

One observes again that the second variation of energy is positive for all paths provided $u_{i,\eta} \neq 0$ and $y_{\delta,\eta} \neq 0$. If

$$u_{i,\eta} = y_{\delta,\eta} = 0$$

higher order variations of Ψ are required. (2.4.84) implies that the only potential candidate for η among the $N+K$ variables is y_1, the critical coordinate. In other words, higher order variations of Ψ with respect to a path which is initially in the form of the ray

$$u_{i,1} = y_{\delta,1} = 0 \tag{2.4.85}$$

must now be examined.

Differentiating (2.4.83) for a third time with respect to $\eta \equiv y_1$ and evaluating at F yields

$$\left.\frac{d^3\psi}{dy_1^3}\right|_F = 0 \tag{2.4.86}$$

indicating an indecisive situation.

Proceeding in the same manner as before, the fourth differentiation yields on evaluation

$$\left.\frac{d^4\psi}{dy_1^4}\right|_F = \Psi_{1111} + 6\Psi_{i11} u_{i,11} + 3\Psi_{ii}(u_{i,11})^2 \tag{2.4.87}$$

2.4.6 Stability distribution on the equilibrium surfaces

From Section 1.1.4, it is deduced immediately that the particular derivative $u_{i,11}$ minimizing (2.4.87) is given by

$$u_{i,11} = -\frac{\Psi_{i11}}{\Psi_{ii}}$$

Substituting for u_{i11} in the equation (2.4.87) one has

$$\left.\frac{d^4\psi}{dy_1^4}\right|_F = \Psi_{1111} - 3\sum_{i=1}^{N}\frac{(\Psi_{i11})^2}{\Psi_{ii}} \qquad (2.4.88)$$

and the stability criterion

$$\Psi_{1111} - 3\sum_{i=1}^{N}\frac{(\Psi_{i11})^2}{\Psi_{ii}} \begin{array}{c} > \\ = \\ < \end{array} 0 \begin{array}{l} \text{stable} \\ \text{for critical equilibrium} \\ \text{unstable} \end{array} \qquad (2.4.89)$$

If in (2.4.89), equality holds, then higher order variations have to be examined, and this can be done readily by following the same procedure.

Comparing (2.4.89) and (2.4.79), one can state the following theorem:

THEOREM 2.15. *The initial post-buckling surface is stable or unstable according to whether the critical point itself is stable or unstable respectively.*

2.5

Examples and experimental results

2.5.1 Four-degree-of-freedom model: symmetric special critical point

To demonstrate some aspects of the general theory concerning special critical points, consider the rigid link model illustrated in Figure 2.35 which consists of two pin-jointed straight members, each rigid in bending but permitting linearly elastic axial deformations (with spring constants $\bar{k}_1$ and $\bar{k}_2$). Each of the two joints has a linear torsional spring with stiffnesses $\bar{k}_3$ and $\bar{k}_4$. Two concentrated loads act vertically at A and B as shown in Figure 2.35.

From the buckled configuration of the system the following geometrical relationships are readily obtained:

$$\Delta\overline{OA} = l - l[z_1^2 + (1-Q_1)^2]^{\frac{1}{2}}$$
$$\Delta\overline{AB} = l - l[(z_2-z_1)^2 + (1+Q_1-Q_2)^2]^{\frac{1}{2}}$$
$$\theta_1 = tg^{-1}\frac{z_1}{1-Q_1} \tag{2.5.1}$$
$$\theta_2 = tg^{-1}\frac{z_2-z_1}{1+Q_1-Q_2} - tg^{-1}\frac{z_1}{1-Q_1}$$

where z_1, z_2, Q_1 and Q_2 are generalized coordinates.

The potential energy V of the system is given by

$$V = \tfrac{1}{2}\bar{k}_1(\Delta\overline{OA})^2 + \tfrac{1}{2}\bar{k}_2(\Delta\overline{AB})^2 + \tfrac{1}{2}\bar{k}_3\theta_1^2$$
$$+ \tfrac{1}{2}\bar{k}_4\theta_2^2 - P_1Q_1l - P_2Q_2l \tag{2.5.2}$$

Using (2.5.1), series expansions and assuming that the Q_i as well as the z_α are small quantities permitting the omission of the terms which will not appear in the first order equilibrium equations, one has the nondimensionalized energy expression

$$V = \tfrac{1}{8}(-2Q_1 + z_1^2)^2 + \tfrac{1}{8}k_2[2(Q_1-Q_2) + (z_2-z_1)^2]^2$$
$$+ \tfrac{1}{2}k_3z_1^2 + \tfrac{1}{2}k_4(z_2-2z_1)^2 - \Lambda^1Q_1 - \Lambda^2Q_2 \tag{2.5.3}$$

195

2.5 Examples and experimental results

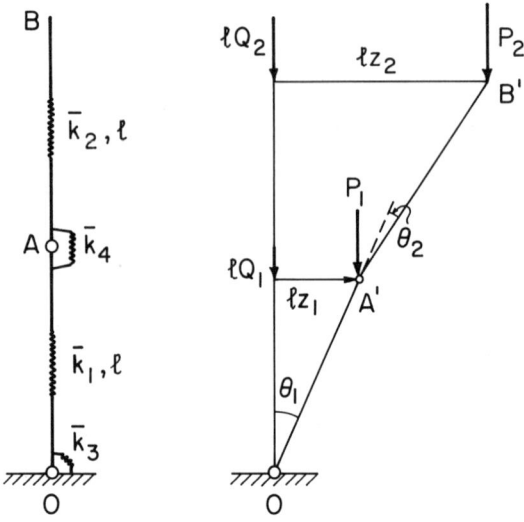

Figure 2.35

where both sides are divided by $\bar{k}_1 l^2$, so that k's are nondimensional and $\Lambda^1 = P_1/\bar{k}_1 l$, $\Lambda^2 = P_2/\bar{k}_1 l$.

Upon comparing (2.5.3) with (2.4.23) one observes that the former is in the form of the latter indicating that the model under consideration falls within the scope of the class treated in Section 2.4.3. It follows that the buckling behaviour of the model must be in compliance with Theorem 2.9.

The equilibrium equations of the system are easily derived as

$$\frac{\partial V}{\partial Q_1} = (1+k_2)Q_1 - k_2 Q_2 - \tfrac{1}{2}z_1^2 + \tfrac{1}{2}k_2(z_2 - z_1)^2 - \Lambda^1 = 0$$

$$\frac{\partial V}{\partial Q_2} = -\tfrac{1}{2}k_2[2(Q_1 - Q_2) + (z_2 - z_1)^2] - \Lambda^2 = 0$$

$$\frac{\partial V}{\partial z_1} = -k_2(Q_1 - Q_2)(z_2 - z_1) - Q_1 z_1 - \tfrac{1}{2}k_2(z_2 - z_1)^3$$
$$+ k_3 z_1 + \tfrac{1}{2}z_1^3 - 2k_4(z_2 - 2z_1) = 0,$$

and

$$\frac{\partial V}{\partial z_2} = k_2(Q_1 - Q_2)(z_2 - z_1) + \tfrac{1}{2}k_2(z_2 - z_1)^3 + k_4(z_2 - 2z_1) = 0$$

which yield the fundamental plane

$$z_1 = z_2 = 0$$
$$(1 + k_2)Q_1 - k_2 Q_2 = \Lambda^1 \qquad (2.5.4)$$
$$-k_2(Q_1 - Q_2) = \Lambda^2$$

2.5.1 Four-degree-of-freedom model: symmetric special critical point

In order to locate the critical points on this plane, proceed to determine the second variation of energy with respect to z_1 and z_2. Second derivatives evaluated for $z_1 = z_2 = 0$ are

$$\left.\frac{\partial^2 V}{\partial z_1^2}\right|_F = k_2(Q_1 - Q_2) - Q_1 + k_3 + 4k_4$$

$$\left.\frac{\partial^2 V}{\partial z_1 \partial z_2}\right|_F = -k_2(Q_1 - Q_2) - 2k_4$$

$$\left.\frac{\partial^2 V}{\partial z_2^2}\right|_F = k_2(Q_1 - Q_2) + k_4$$

Substituting for Q_1 and Q_2 from (2.5.4), one has the determinantal equation

$$\Delta^z \equiv \det \left| \frac{\partial^2 V}{\partial z_\alpha \partial z_\beta} \right|$$

$$\equiv \det \begin{vmatrix} -\Lambda^1 - 2\Lambda^2 + k_3 + 4k_4 & \Lambda^2 - 2k_4 \\ \Lambda^2 - 2k_4 & -\Lambda^2 + k_4 \end{vmatrix} = 0$$

which yields the equation of critical curves (comprising the stability boundary)

$$(\overset{*}{\Lambda}{}^1 + \overset{*}{\Lambda}{}^2)\overset{*}{\Lambda}{}^2 - k_4\overset{*}{\Lambda}{}^1 - (k_3 + 2k_4)\overset{*}{\Lambda}{}^2 + k_3 k_4 = 0 \qquad (2.5.5)$$

On a plot of $\overset{*}{\Lambda}{}^1$ versus $\overset{*}{\Lambda}{}^2$, one gets a hyperbola. Figure 2.36 is drawn

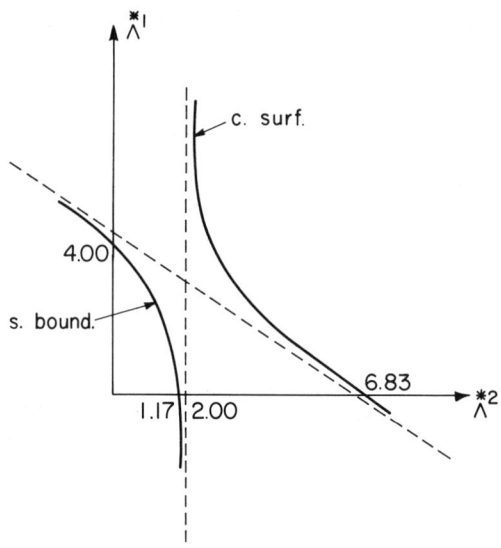

Figure 2.36

2.5 Examples and experimental results

for $k_4 = 2$ and $k_3 = 4$. Clearly, the stability boundary is concave towards the region of stability in compliance with Theorem 2.9.

Problem. Consider an arbitrary ray of the form $\Lambda^i = l^i \xi$ and examine the stability of the corresponding primary critical point. Derive the equations of the post-buckling path in the vicinity of this point. Examine the possible ways of obtaining the post-buckling path corresponding to another ray.

2.5.2 Shallow arch under two independent sets of concentrated loads

The simply supported shallow arch which was considered in Section 1.4.5 will now be analyzed under the combined action of three concentrated loads; one acting at the apex vertically and the other two symmetrically located equal loads acting in radial directions as shown in Figure 2.37.

Following Section 1.4.5, the total potential energy function of this system can be expressed as

$$W = \frac{1}{4}\left(\tfrac{1}{2}c^2 u_1^2 + 2c^2 y_2^2 - \frac{4}{\pi} u_1\right)^2 + d(\tfrac{1}{2}c^4 u_1^2 + 8c^4 y_2^2)$$

$$- \Lambda^1 u_1 - 2\Lambda^2 u_1 \cos\frac{\pi\alpha}{2\theta_0} \qquad (2.5.6)$$

which differs from (1.4.30) in the potential energy of external loads only.

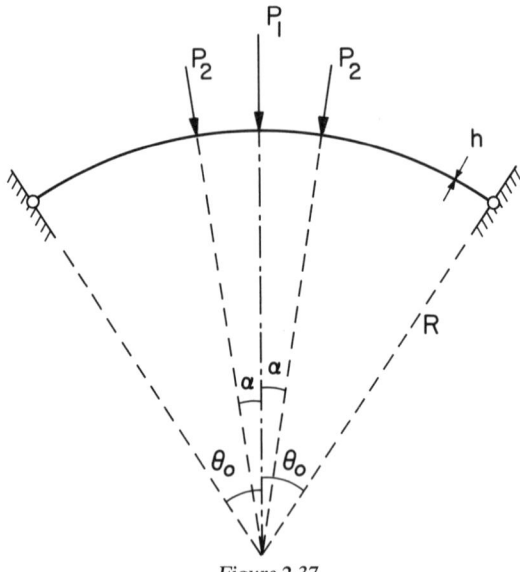

Figure 2.37

2.5.2 Shallow arch under two independent sets of concentrated loads

It is understood that again the deflection function (1.4.29) has been used to reduce the system to one with two degrees of freedom and

$$\Lambda^1 = \frac{P_2}{EA\theta_0} \quad \text{and} \quad \Lambda^2 = \frac{P_2}{EA\theta_0}$$

Consider first a *point by point* analysis as described in Section 2.4.1. The behaviour of the system along the ray $\Lambda^i = l^i \xi$ which is specified by $l^1 = 1$ and $l^2 = 0$ (i.e., when $\Lambda^2 = 0$) is already known from Section 1.4.5, and the bifurcation point, for example, is given by

$$\xi_{cr}^B \equiv \Lambda_B^1 = \frac{4c^2 d}{\pi} + 12c^2 d\left(\frac{1}{\pi^2} - c^4 d\right)^{\frac{1}{2}} \tag{2.5.7}$$

Similarly, the limit points are exactly the same as given in Section 1.4.5.

Now consider the ray which is defined by $l^1 = 0$ and $l^2 = 1$; obviously, the stability equations remain unaffected and from (1.4.33) one has

$$u_1^B = \frac{4}{c^2}\left[\frac{1}{\pi} \pm \left(\frac{1}{\pi^2} - c^4 d\right)^{\frac{1}{2}}\right] \tag{2.5.8a}$$

The fundamental equilibrium path corresponding to this second ray, however, is different and can readily be obtained by substituting for

$$W_1' = -2\cos\frac{\pi\alpha}{2\theta_0}$$

in the equilibrium equations (1.3.14) to (1.3.16). Thus, one has

$$2\cos\frac{\pi\alpha}{2\theta_0}\Lambda^2 = \left(\frac{8}{\pi^2} + dc^4\right)u_1 - \frac{3c^2}{\pi}u_1^2 + \tfrac{1}{4}c^4 u_1^3$$

which after solving with (2.5.8a) simultaneously yields

$$\Lambda_B^2 = \frac{b}{2a} \tag{2.5.8b}$$

where

$$a = \cos\frac{\pi\alpha}{2\theta_0} \quad \text{and} \quad b = \frac{4c^2 d}{\pi} + 12c^2 d\left(\frac{1}{\pi^2} - c^4 d\right)^{\frac{1}{2}}$$

According to Theorem 2.11, the straight line

$$\overset{*}{\Lambda}{}^1 + 2a\overset{*}{\Lambda}{}^2 = b \tag{2.5.9}$$

joining the two critical points (2.5.7) and (2.5.8b) is, then, the stability boundary of the system. In fact, if another ray is considered and the associated critical value $\overset{*}{\xi}$ is determined by solving the corresponding

2.5 Examples and experimental results

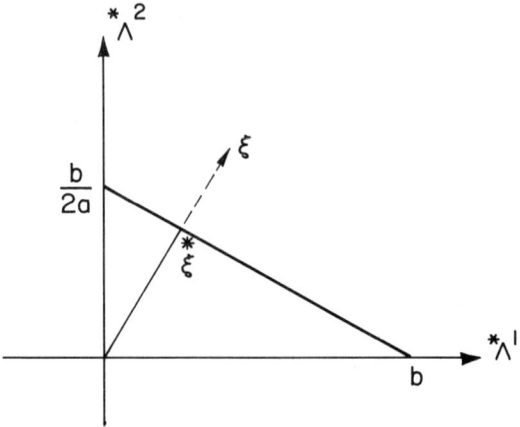

Figure 2.38

equilibrium equation (1.3.16) and (2.5.8a) for ξ one observes that the critical point obtained lies on the line (2.5.9) as illustrated in Figure 2.38.

Problem 1. Depending on the physical properties of the arch, stability may be lost at general critical points; find the associated stability boundary.

Problem 2. Rederive the equation of the stability boundary (2.5.9) associated with special critical points *directly* through a two-parameter perturbation technique (see the next Section).

2.5.3 Shallow arch under uniform pressure and concentrated load at apex

Consider now the same arch subjected to the combined action of uniform pressure and a concentrated vertical load at the apex as shown in Figure 2.39.

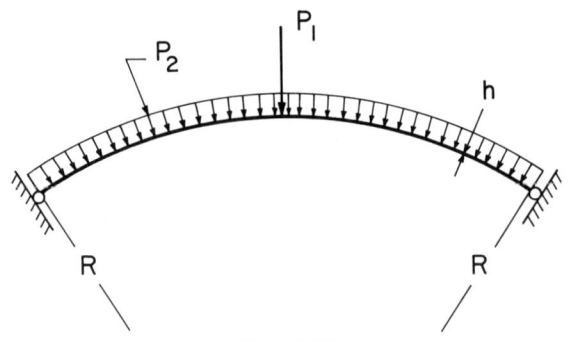

Figure 2.39

2.5.3 Shallow arch under uniform pressure

The total potential energy of this system is similarly obtained as

$$W = \frac{1}{4}\left(c^2 u_1^2 + 2c^2 y_2^2 - \frac{4}{\pi} u_1\right)^2 + d(\tfrac{1}{2} c^4 u_1^2 + 8 c^4 y_2^2)$$

$$- \Lambda^1 u_1 - \frac{4\theta_0 R}{\pi h} \Lambda^2 u_1 \qquad (2.5.10)$$

where $\Lambda^1 = P_1/EA\theta_0$ and $\Lambda^2 = P_2/E\theta_0$.

Here the stability boundary will be estimated *as a whole* rather than *point by point*.

For illustration, ignore the symmetry properties of the system under consideration for now deliberately and express the fundamental surface in the form

$$\begin{aligned}\Lambda^1 &= \Lambda^1(\eta^1, \eta^2) \\ \Lambda^2 &= \Lambda^2(\eta^1, \eta^2) \\ u_1 &= u_1(\eta^1, \eta^2) \\ y_2 &= y_2(\eta^1, \eta^2)\end{aligned} \qquad (2.5.11)$$

although it is very well known that $y_2 = 0$ along the fundamental surface.

To apply the general perturbation technique described in Section 2.2.1, evaluate first the energy derivatives at the origin. Although most of these derivatives are available in Section 1.4.5, they are produced here as a complete set for convenience:

$$W_{11} = \frac{8}{\pi^2} + dc^4, \qquad W_{22} = 16\, dc^4$$

$$W_{111} = -\frac{6c^2}{\pi}, \qquad W_{122} = -\frac{8c^2}{\pi} \qquad (2.5.12)$$

$$W_{1111} = \tfrac{3}{2} c^4, \qquad W_{1122} = 2c^4, \qquad W_{2222} = 24 c^4$$

$$W_1^1 = -1, \qquad W_2^1 = 0, \qquad W_1^2 = -\frac{4\theta_0 R}{\pi h}, \qquad W_2^2 = 0$$

The first order perturbation equations (2.2.4) then, take the form

$$\begin{aligned}&\left(\frac{8}{\pi^2} + dc^4\right) u_1^1 - \Lambda^{1,1} - \frac{4\theta_0 R}{\pi h} \Lambda^{2,1} = 0 \quad \text{for} \quad i = 1 \\ & 16\, dc^4 y_2^1 \qquad\qquad\qquad\qquad\qquad\qquad \text{for} \quad i = 2\end{aligned} \qquad (2.5.13)$$

and

$$\begin{aligned}&\left(\frac{8}{\pi^2} + dc^4\right) u_1^2 - \Lambda^{1,2} - \frac{4\theta_0 R}{\pi h} \Lambda^{2,2} = 0 \quad \text{for} \quad i = 1 \\ & 16\, dc^4 y_2^2 = 0 \qquad\qquad\qquad\qquad\qquad\qquad \text{for} \quad i = 2\end{aligned} \qquad (2.5.14)$$

2.5 Examples and experimental results

From (2.5.13) and (2.5.14) one first obtains

$$y_2^1 = y_2^2 = 0 \qquad (2.5.15)$$

which indicates clearly that neither η^1 nor η^2 can be taken as y_2, i.e., the analysis itself rules out the possibility of y_2 wrongly being chosen as a perturbation parameter. This is an important feature of the technique, because the selection of the independent parameters does not heavily depend on the ability of the analyzer.

Suppose one sets $\eta^1 \equiv u_1$ and $\eta^2 = \Lambda^2$ (Λ^1 can also be chosen), then

$$\Lambda^{2,1} \equiv \Lambda_1^2 \equiv \partial \Lambda^2 / \partial u_1 = 0,$$
$$u_1^2 \equiv \partial u_1 / \partial \Lambda^2 = 0$$

since u_1 and Λ^2 are now independent parameters, and the equations (2.5.13) to (2.5.14) yield the surface derivatives

$$\Lambda_1^1 = \frac{8}{\pi^2} + dc^4$$

and (2.5.16)

$$\Lambda^{1,2} = -\frac{4\theta_0 R}{\pi h}$$

respectively.

The second order equations (2.2.5) yield for $i = 1$, $m = 1$, $l = 1$,

$$W_{111} u_{1,1} u_{1,1} + W_{122} y_{2,1} y_{2,1} + W_1^1 \Lambda_{11}^1 = 0$$

which upon using (2.5.15) results in

$$\Lambda_{11}^1 = -\frac{6c^2}{\pi} \qquad (2.5.17)$$

Similarly, from equations (2.2.5) one has

$\Lambda^{1,22} = 0$ for $i = 1, m = 2, l = 2$
$\Lambda_1^{1,2} = 0$ for $i = 1, m = 1, l = 2$
$y_{2,11} = 0$ for $i = 2, m = 1, l = 1$ (2.5.18)
$y_2^{22} = 0$ for $i = 2, m = 2, l = 2$
$y_{2,1}^2 = 0$ for $i = 2, m = 2, l = 1$

The third perturbation yields

$$\Lambda_{111}^1 = \tfrac{3}{2} c^4 \qquad (2.5.19)$$

The ordered equations of the fundamental equilibrium surface can

2.5.3 Shallow arch under uniform pressure

now be constructed as

$$\Lambda^1 = \Lambda^1(u_1, \Lambda^2) = \left(\frac{8}{\pi^2} + dc^4\right)u_1 - \frac{4\theta_0 R}{\pi h}\Lambda^2$$

$$y_2 = y_2(u_1, \Lambda^2) = 0$$

$$\Lambda^1 = \left(\frac{8}{\pi^2} + dc^4\right)u_1 - \frac{4\theta_0 R}{\pi h}\Lambda^2 - 3\frac{c^2}{\pi}u_1^2$$

$$y_2 = 0$$

and

$$\Lambda^1 = \left(\frac{8}{\pi^2} + dc^4\right)u_1 - \frac{4\theta_0 R}{\pi h}\Lambda^2 - 3\frac{c^2}{\pi}u_1^2 + \tfrac{1}{4}c^4 u_1^3$$

$$y_2 = 0. \tag{2.5.20}$$

The third order equation (2.5.20) of the fundamental surface could, of course, be obtained directly in this two-degree-of-freedom analysis. However, when one considers higher degrees of freedom, such a direct approach becomes unmanageable while the illustrated perturbation procedure preserves its essential systematic character. It must also be noted that, if the obvious symmetry properties of the system which results in $y_2 = 0$ along the fundamental surface were utilized in the beginning of the analysis, the construction of (2.5.20) would have been much simpler.

In order to locate the critical points on the fundamental equilibrium surface, one examines the variations of the stability determinant

$$\Delta(u_1, y_2) \equiv \det |W_{uy}(u_1, y_2)| \tag{2.5.21}$$

along this surface either by ignoring or taking into account the fact that (2.5.21) can be expressed as a product of two distinct determinants Δ^u and Δ^y.

In the latter approach, one can use the formula-type results (1.3.23) to (1.3.25) for *general* critical points and (1.3.29) for *special* critical points to obtain

$$A_0^u = \frac{8}{\pi^2} + dc^4, \qquad A_0^y = 16\, dc^4$$

$$A_1^u = -\frac{6c^2}{\pi}, \qquad A_1^y = -8\frac{c^2}{\pi} \tag{2.5.22}$$

$$A_{11}^u = \tfrac{3}{2}c^4, \qquad A_{11}^y = 2c^4$$

just like in Section 1.4.5. These derivatives, then, yield the second order stability equations

$$\frac{8}{\pi^2} + dc^4 - \frac{6c^2}{\pi}u_1 + \tfrac{3}{4}c^4 u_1^2 = 0 \tag{2.5.23}$$

203

2.5 Examples and experimental results

and

$$16\,dc^4 - 8\frac{c^2}{\pi}u_1 + c^4 u_1^2 = 0 \tag{2.5.24}$$

for general and special critical points respectively.

Solving (2.5.20) with (2.5.24), for example, concurrently, one has the stability boundary associated with special critical points as

$$\overset{*}{\Lambda}{}^1 + \frac{4\theta_0}{\pi h}\overset{*}{\Lambda}{}^2 = b \tag{2.5.25}$$

where b is given in (2.5.8).

For comparison, consider next an analysis which ignores the fact that the stability determinant (2.5.21) is the product of Δ^u and Δ^y. Then, one cannot use the formula-type results (1.3.23) to (1.3.25) or (1.3.29), and computations become lengthier. Thus, differentiating (2.5.21) with respect to appropriate variables and evaluating result in

$$\Delta_1 = -\frac{96}{\pi}dc^6 - 8\left(\frac{8}{\pi^2} + dc^4\right)\frac{c^2}{\pi}$$

$$\Delta_2 = \Delta^1 = \Delta^2 = 0$$

$$\Delta_{11} = 24\,dc^8 + 96\frac{c^4}{\pi^2} + 2c^4\left(\frac{8}{\pi^2} + dc^4\right) \tag{2.5.26}$$

$$\Delta_{12} = 0$$

$$\Delta_{22} = 32\,dc^8 - 128\frac{c^4}{\pi^2} + 24c^4\left(\frac{8}{\pi^2} + dc^4\right),$$

etc.

Note that any derivative with respect to loading parameters vanishes since these parameters do not appear in the determinant. Now, the surface derivatives (2.5.16) to (2.5.19) of the equilibrium analysis can be used to evaluate A^k, A^{kl}, etc. in equations (2.2.12). Thus,

$$A_1 = \Delta_1 u_{1,1} + \Delta_2 u_{2,1} + \Delta^1 \Lambda_1^1$$

leads to

$$A_1 = \Delta_1.$$

Similarly

$$A^2 = \Delta_1 u_1^2 + \Delta_2 y_2^2 + \Delta^1 \Lambda^{1,2} + \Delta^2$$

leads to

$$A^2 = 0.$$

2.5.3 Shallow arch under uniform pressure

Furthermore, one has

$A_{11} = \Delta_{11}$

$A_1^2 = A^{22} = 0$

$A_{111} = \Delta_{111} = -72\dfrac{c^6}{\pi}$

$A_{1111} = \Delta_{1111} = 18c^8$

and other derivatives are all zero. Noting that

$$A_0(0,0) = \Delta_0 = 16\left(\dfrac{8}{\pi^2} + dc^4\right) dc^4$$

one can construct the ordered stability equations up to fourth order. The latter equation is in the form

$$\Delta_0 + \Delta_1 u_1 + \tfrac{1}{2}\Delta_{11} u_1^2 + \dfrac{1}{3!}\Delta_{111} u_1^3 + \dfrac{1}{4!}\Delta_{1111} u_1^4 = 0 \qquad (2.5.27)$$

which after substitutions can be written as

$$\left(\dfrac{8}{\pi^2} + dc^4 - \dfrac{6c^2}{\pi} u_1 + \tfrac{3}{4}c^4 u_1^2\right)\left(16\, dc^4 - 8\dfrac{c^2}{\pi} u_1 + c^4 u_1^2\right) = 0 \qquad (2.5.28)$$

Depending on the physical properties of the arch either the expression in the first or second parentheses vanishes, giving the critical values of u_1 associated with *general* or *special* critical points on the surface, respectively. These two conditions are, of course, recognized as precisely those given in (2.5.23) and (2.5.24), and one realizes that ignoring the symmetry properties of the system necessitated a fourth order stability equation instead of a second order one for the same order approximation.

By truncating the left-hand side of (2.5.27) after the second, third, etc., terms, one obtains first, second, etc., order stability equations which normally differ from the ordered estimates associated with Δ^u and/or Δ^y individually.

Solving (2.5.28) with (2.5.20) concurrently yields the stability boundaries associated with general and special critical points and the latter, for example, is again in the form (2.5.25).

Problem. Introduce the necessary transformations to obtain the function $\Psi(u_i, y_\alpha, \Phi^j)$ from the function W, and use the expressions in Section 2.4.4, to construct the first order equations of the equilibrium surface. Discuss the stability of this surface (see Ref. [48]).

2.5 Examples and experimental results

2.5.4 Experimental results

A series of experiments was performed on shallow arches under the combined action of certain symmetrically located sets of concentrated loads [43].

The particular test reported here was performed on a shallow circular arch made of high tensile steel strip with dimensions $1 \times \frac{1}{16}$ in. The arch was rolled to form an approximately circular curve with a span of 24 inches and a rise of 1·30 in. The simply supported edge conditions were obtained by filing the ends of the arch into knife-edges and fitting them into fixed horizontal V-grooves.

The arch was subjected to the combined action of three symmetrically located loads, one acting at the apex and the other two (equal loads) acting on either side of the apex at equal distances as shown in Figure 2.40a. The two symmetrically applied side loads Λ^2 and the central load Λ^1 were changed independently. One of the sets was increased gradually while the other was kept constant until buckling occurred. It was observed that the loss of stability was always associated with asymmetric deflections. In other words, under all combinations of Λ^1 and Λ^2, instability took place at a point of bifurcation where the initially symmetric deflections gave way to asymmetric deflections. The results of the test are shown in Figure 2.40b where Λ^1_{cr} and Λ^2_{cr} are the critical values of the

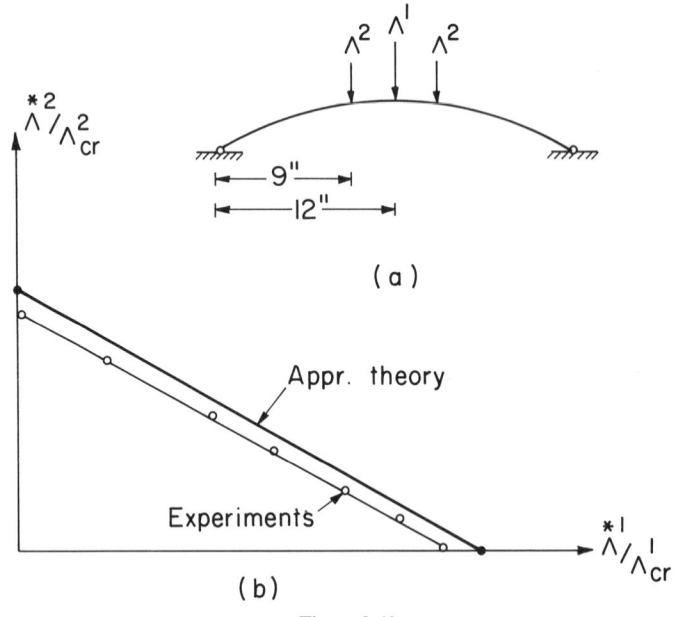

Figure 2.40

2.5.4 Experimental results

corresponding parametes obtained in the absence of each other. It is seen that all the experimental points lie approximately on a straight line as Theorem 2.11 states. Apparently, the system behaves very much like a two-degree-of-freedom one.

For a quantitative comparison, the stability boundary can be determined theoretically by means of (2.5.9). Since the angle α is very small ($\alpha = 0.053$) only the radial components of the vertical side loads may be considered. The result of such an analysis is in good agreement with the exeriments as Figure 2.40b indicates.

Concluding remarks

A general nonlinear theory of elastic stability for conservative (statical) systems has been presented. Although the book removes one of the traditional limitations associated with previous treatments of the subject by exposing the multiple-parameter systems systematically, nevertheless, it does not touch on the fascinating field of nonconservative forces, neither does it offer a clue as to the inelastic behaviour. Similarly, dynamic loading conditions, thermodynamics of elastic stability and stochastic effects have been omitted. In view of the rapidly growing literature concerning all aspects of stability theory, however, this must be interpreted as a purposely and perhaps inevitable confinement.

The energy criterion which forms the basis of the analyses throughout this book has to be applied with extreme care once the domain of the present subject is left. Even for *gyroscopic conservative* systems a relative minimum of the potential energy ceases to be a necessary condition for stability although it remains as a sufficient one, and these systems may lose stability by *flutter* as well as by divergence [99].

On the other hand, there exists a distinct class of nonconservative elastic systems whose behaviour is analogous to that of conservative systems: frequencies of free vibrations are always real and a loss of stability only occurs via divergence. Such systems may be called *pseudoconservative* and they can be analyzed through statical criteria. Instability of multiple-parameter pseudo-conservative systems has recently been discussed in general terms [89].

Finally, one has strongly nonconservative elastic systems which can lose stability by flutter as well as by divergence. General discussions concerning the stability of such systems can be found in the excellent books by Bolotin, Ziegler and Leipholz listed at the end of the references. Recently, the stability of a two-parameter linear nonconservative system has been discussed in Ref. [100]. It has been shown that the stability boundary, which consists of flutter and divergence boundaries, is convex towards the origin as opposed to what Theorem 2.9 established herein states for linear conservative systems. This is of great significance in obtaining estimates of the stability boundary.

References

[1] POINCARÉ, H., Sur l'équilibre d'une masse fluide animée d'un mouvement de rotation. *Acta Math. Stockh.* 7 (1885) 259.
[2] BRYAN, G. H., On the stability of elastic systems. *Proc. Cambr. Phil. Soc.* 6 (1888) 199.
[3] SOUTHWELL, R. V., On the general theory of elastic stability. *Phil. Trans. Roy. Soc. London, Ser. A, 213* (1913) 187.
[4] BIEZENO, C. B. and HENCKY, H., On the general theory of elastic stability. *Proc. Acad. Sci. Amsterdam,* 31 (1928) 569, 32 (1929) 444.
[5] REISSNER, H., Energy criterion of resistance to buckling. *Z. Angew. Math. Mech.,* 5 (1925) 475.
[6] TREFFTZ, E., Derivation of stability criteria of elastic equilibrium from the elasticity theory of finite deformations. *Proc. Third Intern. Congr. Appl. Mech., Stockholm 3* (1930) 44.
[7] TREFFTZ, E., Theory of stability of the elastic equilibrium. *Z. Angew. Math. Mech.* 13 (1933) 160.
[8] MARGUERRE, K., Treatment of stability problems by means of the energetic method. *Z. Angew. Math. Mech.* 18 (1938) 57.
[9] KAPPUS, R., Theory of elasticity of finite displacements. *Z. Angew. Math. Mech.* 19 (1939) 271–344.
[10] BIOT, M., Theory of elasticity with large displacements and rotations. *Proc. Fifth Intern. Congr. Appl. Mech.,* p. 117, Cambridge, Mass. (1938).
[11] BIOT, M., Theory of elasticity of the second order, with practical applications. *Z. Angew. Math. Mech.* 20 (1940) 89.
[12] KOITER, W. T., On the stability of elastic equilibrium, Dissertation, Delft, Holland (1945). English translation, NASA, TT, F10 (1967) 833.
[13] KOITER, W. T., Stability of equilibrium of continuous bodies. Brown University, TR No. 79, April (1962).
[14] KOITER, W. T., On the concept of stability of equilibrium for continuous bodies. *Proc. R. Neth. Acad. Sci., Ser.* [B]66 (1963) 173–177.
[15] KOITER, W. T., Elastic stability and post-buckling behaviour. *Proc. Symp. Nonlinear Problems,* edited by R. E. Langer, p. 257. University of Wisconsin Press (1963).
[16] KOITER, W. T., On the instability of equilibrium in the absence of a minimum of the potential energy. *Proc. R. Neth. Acad. Sci., Ser.* [B]68 (1965) 107–113.
[17] KOITER, W. T., The energy criterion of stability for continuous elastic bodies. *Proc. R. Neth. Acad. Sci., Ser.* [B]68 (1965) 178–202.
[18] KOITER, W. T., Purpose and achievements of research in elastic stability. *Proc. Fourth Technical Conference, Soc. for Engng. Sci.,* North Carolina State University, Rayleigh, N.C., Nov. (1966).

References

[19] KARMAN, TH. VON, and TSIEN, H. S., The buckling of thin cylindrical shells under axial compression. *J. Aero. Sci.* 8 (1941) 303.
[20] BUDIANSKY, B. and HUTCHINSON, J. W., Dynamic buckling of imperfection-sensitive structures. *Proc. XI Internat. Congr. Appl. Mech.*, Munich, pp. 636–651. Julius Springer-Verlag, Berlin (1964).
[21] BUDIANSKY, B., Dynamic buckling of elastic structures: criteria and estimates. *Dynamic Stability of Structures, Proc. of an International Conference held at Northwestern University, Evanston, Illinois*, Pergamon Press, Oct. (1965).
[22] BUDIANSKY, B. and HUTCHINSON, J. W., A survey of some buckling problems. *AIAA Journal* 4 (1966) 1505–1510.
[23] HUTCHINSON, J. W. and BUDIANSKY, B., Dynamic buckling estimates. *AIAA Journal* 4 (1966) 525–530.
[24] HUTCHINSON, J. W., Imperfection-sensitivity of externally pressurized spherical shells. *J. Appl. Mech.* 34 (1967) 49–55.
[25] HUTCHINSON, J. W., Initial post-buckling behaviour of toroidal shell segments. *Int. J. Solids Struct.* 3 (1967) 97–115.
[26] HUTCHINSON, J. W., AMAZIGO, J. C., Imperfection-sensitivity of eccentrically stiffened cylindrical shells. *AIAA Journal* 5 (1967) 392–401.
[27] BRITVEC, S. J. and CHILVER, A. H., Elastic buckling of rigidly-jointed braced frames. *J. Engng. Mech. Div. Am. Soc. Civ. Engrs.*, 89 (1963) 217.
[28] GODLEY, M. H. R., Elastic buckling of plane frames, Ph.D. Thesis, University College, London (1966).
[29] HAYMAN, B., The stability of degenerate structures, Ph.D. Thesis, University College London (1970).
[30] PERRY, S. H., Statistical variation of buckling strength, Ph.D. Thesis, University College London (1966).
[31] THOMPSON, J. M. T. and WALKER, A. C., A general theory for the branching analysis of discrete structural systems. *Int. J. Solids Structures*, 5 (1969) 281.
[32] BURGESS, I. W., On the equilibrium and stability of discrete one-way structural systems. *Int. J. Solids Structures*, 7 (1971) 667.
[33] CHETAYEV, N. G., *The stability of motion*. Translated from the Russian second edition (1955) by M. Nadler. Pergamon Press, Oxford, London, New York, Paris (1961).
[34] SOKOLNIKOFF, I. S., *Tensor analysis*. John Wiley and Sons, Inc. (1964).
[35] THOMPSON, J. M. T., Basic principles in the general theory of elastic stability. *J. Mech. Phys. Solids* 11 (1963) 13–20.
[36] ROORDA, J., The instability of imperfect elastic structures, Ph.D. Thesis, University College, London (1965).
[37] VAN DYKE, M. D., *Perturbation methods in fluid mechanics*. Academic Press (1964).
[38] BELLMANN, R. E., *Perturbation techniques in mathematics, physics and engineering* Holt, Rinehart and Winston (1964).
[39] COLE, J. D., *Perturbation methods in applied mathematics*. Blaisdell (1968).
[40] SEWELL, M. J., The static perturbation technique in buckling problems. *J. Mech. Phys. Solids* 13 (1965) 247–265.
[41] THOMPSON, J. M. T., Discrete branching points in the general theory of elastic stability. *J. Mech. Phys. Solids* 13 (1965) 295–310.
[42] SEWELL, M. J., On the connexion between stability and the shape of the equilibrium surface. *J. Mech. Phys. Solids* 4 (1966) 203–230.
[43] HUSEYIN, K., Elastic stability of structures under combined loading. Ph.D. Thesis, University College, London University, England, Submitted in December (1967).
[44] HUSEYIN, K., The multiple parameter perturbation technique for the analysis of

References

non-linear systems. *Int. J. of Non-Linear Mechanics* 8 (1973) 431–443. See also SMD Report No. 69, University of Waterloo (1971).

[45] HUSEYIN, K., The elastic stability of structural systems with independent loading parameters. *Int. J. of Solids and Structures* 6 (1970) 677–691. See also Mechanics, American Academy of Mechanics 1 (1970) 183–199.

[46] HUSEYIN, K., The stability boundary of systems with one-degree-of-freedom, Part I. *Meccanica* 5 (1970) 308–311.

[47] HUSEYIN, K., The stability boundary of systems with one-degree-of-freedom, Part II. *Meccanica* 5 (1970) 312–316.

[48] HUSEYIN, K., Instability of symmetric structural systems with independent loading parameters. *Quart. Appl. Math.* 28 (1971) 571–586.

[49] HUSEYIN, K., The post-buckling behaviour of structures under combined loading. *Z. Angew. Math. Mech.* 51 (1971) 177–182.

[50] HUSEYIN, K., On the estimation of the stability boundary of symmetric structural systems. *J. of Non-linear Mech.* 7 (1972) 31–50. See also SMD Report No. 34, U. of Waterloo (1971).

[51] HUSEYIN, K., Singular critical points in the general theory of elastic stability. *Meccanica* 7 (1972) 58–68. See also SMD Report No. 74, U. of Waterloo (1971).

[52] HUSEYIN, K., On the stability of critical equilibrium states and post-buckling, Part 1, Discrete systems, Part II. Continuous systems, study No. 6, Stability, edited by H. Leipholz, U. of Waterloo, pp. 15–40 (1972).

[53] HUSEYIN, K., On the stability of critical equilibrium states, SMD Report No. 61, U. of Waterloo (1970).

[54] THOMPSON, J. M. T., The estimation of elastic critical loads. *J. Mech. Phys. Solids* 15 (1967) 311–317.

[55] MASUR, E. F. and SCHREYER, H. L., A second approximation to the problem of elastic instability. *Proc. Symp. in Honour of Lloyd H. Donnell*, edited by M. Muster, University of Houston, April (1966).

[56] DANIELSON, D. A., Buckling and initial post-buckling behaviour of spheroidal shells under pressure. Report SM-22, Harvard, March (1968).

[57] ROORDA, J., Stability of structures with small imperfections. *J. Eng. Mech. Div. ASCE* 91 No. EM1 (1965) 87–196.

[58] ROORDA, J., The buckling behaviour of imperfect structural systems, *J. Mech. Phys. Solids* 13 (1965) 267–280.

[59] ROORDA, J., On the buckling of symmetric structural systems with first and second order imperfections, *Int. J. Solids Struct.* 4 (1968) 1137–1148.

[60] WESTERDGAARD, H. M., Buckling of elastic structures, *Trans. Amer. Soc. Engrs.*, 85 (1922) 576.

[61] SALET, G., La théorie générale du flambement élastique de Thuloup, exposé, précisions compléments. *Bull. Ass., Tech. Merit. Aeronaut.* 64 (1964) 205.

[62] HUSEYIN, K., The convexity of the stability boundary of symmetric structural systems. *Acta Mech.* 8 (1969) 205–211.

[63] MASUR, E. F., Buckling, postbuckling and limit analysis of completely symmetric elastic structures, *Int. J. Solids Struct.* 6 (1970) 587–604.

[64] TIMOSHENKO, S. P. and S. WOINOWSKY-KRIEGER. *Theory of plates and shells*, 2nd ed., New York. McGraw-Hill (1959).

[65] SCHREYER, H. L., and MASUR, E. F., Buckling of shallow arches, *J. Engr. Mech. Div., ASCE*, 92, EM4 (1966) 1–19.

[66] GJELSVICK, A. and BODNER, S. R., "The energy criterion and snapbuckling of arches", *A.S.C.E. Proc., Journal of the Eng. Mechanics Division*, Vol. 88, No. 5 (1962).

References

[67] KOITER W. T., The effect of axisymmetric imperfections on the buckling of cylindrical shells under axial compressions. *Proc. R. Neth. Acad. Sci., Ser.* [B]66 (1963) 265.
[68] STEIN, M., The phenomenon of change in buckle pattern in elastic structures. *NASA TR R-39* (1969).
[69] AUGUSTI, G., Stabilità di strutture elastiche elementari in presenza di grandi spostamenti. Publication No. 172, Università di Napoli, Naples (1964).
[70] CHILVER, A. H., Coupled modes of elastic buckling. *J. Mech. Phys. Solids*, 15 (1967) 15-28.
[71] SUPPLE, W. J., Coupled buckling modes of structures. Ph.D. Thesis, University College London (1966).
[72] SUPPLE, W. J., "Coupled branching configurations in the elastic buckling of symmetric structural systems. *Int. J. Mech., Sci.* 9 (1967) 97.
[73] CHILVER, A. H. and JOHNS, K. C., Coincident branching points as generators of buckling paths. *Proc. Cancam Conf.*, Waterloo (1969).
[74] JOHNS, K. C., Elastic buckling behaviour at coincident branching points. Ph.D. Thesis, University College, London (1970).
[75] JOHNS, K. C. and CHILVER, A. H., Multiple path generation at coincident branching points. Tech. Report No. KJ-1 (1971).
[76] SEWELL, M. J., On the branching of equilibrium paths, *Proc. Roy. Soc. Lond. A 315* (1970) 499-518.
[77] THOMPSON, J. M. T. and HUNT, G. W., A theory for the numerical analysis of compound branching, *J. of. Appl. Math. and Phys.*, Vol. 22, Fasc. 6 (1971).
[78] SEWELL, M. J., A general theory of equilibrium paths through critical points, Part I, *Proc. Roy. Soc. A.306* (1968) 201-223; Part II, *Proc. Roy. Soc. A.306* (1968) 225-238.
[79] HO, D., The Influence of imperfections on systems with coincident buckling loads. *Int. J. Nonlinear Mech.* 7 (1972) 311-321.
[80] PAPKOVICH, P. F., *Works on the structural mechanics of ships*. Vol 4 (1963).
[81] RENTON, J. D., *Thin walled structures*. Edited by A. H. Chilver, pp. 1-59, London: Chatto and Windus (1967).
[82] BUCKENS, F., Über Eigenwertscharen, *Öst. Ing. Arch.* 12 (1958) 82-93.
[83] HUSEYIN, K. and ROORDA, J., The loading frequency relationship in multiple eigenvalue problems, *J. Appl. Mech.* 38 (1971) 1007-1011. See also SMD Report 74, U. of Waterloo (1971).
[84] HUSEYIN, K., Fundamental principles in the buckling of structures under combined loading. *Int. J. Solids Struct.* 6 (1970) 479-487.
[85] BIRKHOFF, G. and MACLANE, S., *A survey of modern algebra*. Third ed. The Macmillan Co., (1965).
[86] LOO, T. C. and EVAN-IVANOWSKI, R. M., Interaction of critical pressures and critical concentrated loads acting on shallow spherical shells. *J. Appl. Mech.* 33 (1966).
[87] HUSEYIN, K. and PLAUT, R., Extremum properties of generalized Rayleigh Quotient associated with flutter instability. *Quart. Appl. Math.* 32 (1974) 189-201. See also SMD Report No. 112, U. of Waterloo (1972).
[88] HUSEYIN, K. and PLAUT, R., Application of Rayleigh Quotient to eigenvalue problems of pseudo-conservative systems. *J. Sound and Vibrations* 33 (1974) 201-210.
[89] HUSEYIN, K. and LEIPHOLZ, H., Divergence instability of multiple-parameter circulatory systems, *Quart. Appl. Math. 31* (1973) 185-197. See also SMD Report No. 69, U. of Waterloo (1971).
[90] LEIPHOLZ, H. and MASUR, E. F., Discussion on the paper "The loading frequency

References

relationship in multiple eigenvalue problems" and closure. *J. Appl. Mech.* **39** (1972) 636.

Related Articles

[91] ZIEGLER, H., On the concept of elastic stability. *Advances in Applied Mechanics* **4** (1956) 351.
[92] PEARSON, C. E., General theory of elastic stability. *Quart. Appl. Math.* **14** (1956) 133.
[93] HILL, R., On uniqueness and stability in the theory of finite elastic strain. *J. Mech. Phys. Solids* **5** (1957) 229–241.
[94] SHIELD, R. T. and GREEN, A. E., On certain methods in the stability theory of continuous systems. *Arch. Rat. Mech. Anal.* **12** (1963) 354.
[95] SHIELD, R. T., On the stability of linear continuous systems. *Z. Angew. Math. Phys.* **16** (1965) 649–686.
[96] KNOPS, R. J., and WILKES, E. W., On Movchan's theorems for stability of continuous systems. *Int. J. Eng. Sci.* **4** (1966) 303–329.
[97] HUTCHINSON, J. W. and KOITER, W. T., Post-buckling theory. *Appl. Mech. Rev.*, **9** (1971) 1353.
[98] COLLATZ, L., Mehrparametrige Eigenwertaufgaben in unitaren Raumen. *Bulletin Math.* **12(60)** (1968) 41–56.
[99] HUSEYIN, K. and PLAUT, R., Transverse vibrations and stability of systems with gyroscopic forces. *J. Struct. Mech*, to appear.
[100] HUSEYIN, K. and PLAUT, R., The elastic stability of two-parameter non-conservative systems. *J. Appl. Mech.* **40-1**, Series E, (1973) 175–180.
[101] LEIPHOLZ, H., Über ein Kriterium für die Gültigheit der statischen Methode zur Bestimmung der Knichlast von elastischen Staben unter nichtkonservativer Belastung. *Ing. Arch.* **32** (1963) 286–296.
[102] HERRMANN, G., Stability of equilibrium of elastic systems subjected to non-conservative forces. *Appl. Mech. Rev.* **20** (1967) 103–108.
[103] NEMAT-NASSER, S., Thermoelastic stability under general loads. *Appl. Mech. Rev.* **23** (1970) 615–625.
[104] ARIARATHAM, S. T., Stability of structures under stochastic disturbances, *IUTAM Symposium, Herrenalb 1969*, Edited by Leipholz, H., Springer-Verlag (1971).
[105] PLAUT, R. and HUSEYIN, K., Derivatives of eigenvalues and eigenvectors in non-self adjoint systems. *AIAA Journal* **11** (1973) 250.

Books

[1] BLEICH, F., *Buckling strength of metal structures*. McGraw-Hill (1952).
[2] TIMOSHENKO, S. P. and GERE, J. M., *Theory of elastic stability*, McGraw-Hill (1961).
[3] GERARD, G., *Introduction to structural stability theory*. McGraw-Hill (1962).
[4] PFLÜGER, A., *Stabilitätsprobleme der Elastostatik*. Springer-Verlag, 2nd ed., (1964).
[5] VOLMIR, A. S., *Stability of deformable systems*. NASA, AD 628508, (1965).
[6] BOLOTIN, V. V., *Non-conservative problems of the theory of elastic stability*. Pergamon Press (1965).
[7] GREGORY, M. S., Elastic instability: *Analysis of buckling modes and loads of framed structures*. Spon, London (1967).
[8] ZIEGLER, H., *Principles of structural stability*. Blaisdell (1968).
[9] LEIPHOLZ, H., *Stability theory*. Academic Press (1970).
[10] CROLL, J. G. A., and WALKER, A. C., *Elements of structural stability*. Macmillan (1972).
[11] BRITVEC, S. J., *The stability of elastic systems*. Pergamon Unified Eng. Series, Vol. 12 (1973).

References

[12] LEIPHOLZ, H., *Über die direkte Methode der Variationsrechnung zur Bestimmung von Eigenwerten technischer Probleme*, Braun Verlag (1974).
[13] SUPPLE, W. J. (Ed.) *Structural instability*, IPC Science and Technology Press Ltd. (1973).
[14] THOMPSON, J. M. T. and HUNT, G. W., *A general theory of elastic stability*, John Wiley & Sons (1973).
[15] ROORDA, J. *Concepts in elastic structural stability*, Mechanics today (ed. by N. Nasser) vol. 1, Pergamon (1972), 322–372.

Index

A

AMAZIGO, J. C., xviii
AUGUSTI, G., 79

B

BELLMAN, R. E., 14
BIEZENO, C. B., xvii
Bifurcation points, xvi, 7, 12, 25, 26, 36, 44, 45, 56, 75, 97, 99, 100, 118
   A symmetric point of bifurcation, xvii, 13, 25, 26, 45, 61, 64
   Stable symmetric point of bifurcation, xvii, 13, 25, 26, 30, 45, 61, 67
   Unstable symmetric point of bifurcation, xvii, 13, 25, 26, 30, 45, 61, 70
BIOT, M., xvii
BIRKHOFF, G., 127
BODNER, S. R., 70
BOLOTIN, V. V., 209
Branching points (*see* Bifurcation points)
BRITVEC, S. J., xviii
BRYAN, G. H., xvii
BUCKENS, F., 109
Buckling, 49
   deformations, 48
   in-plane, 48
   out-of-plane, 48
   post-buckling, 13, 21, 28, 29, 61, 65, 76, 87, 163
   pre-buckling, 14, 21, 109
   simultaneous, 79, 80
   snap-buckling, 12, 155, 156

BUDIANSKY, B., xviii
BURGESS, I. W., xix

C

CHETAYEV, N. G., xxii
CHILVER, A. H., xviii, 79, 92, 101
Classification of conservative systems, xviii, xix
COLE, J. D., 14
Compound branching, xix, 79, 86 (*see also* Bifurcation points)
Configuration space (*see* Load-deflection space)
Conservative systems, xviii
Continuous systems, xix, xx, xxi, 33
Convexity
   energy surface, 17, 80
   stability boundary, 123, 134, 136, 138, 140, 157, 158, 170, 176, 188
   strict, 17, 134
Critical curves, 197
Critical line, 45, 58, 59, 77
Critical loads, 20, 36, 39, 40, 41, 42, 45, 50, 53, 54, 60, 63, 69, 98, 101
Critical points (states), xvi, 6, 13, 26, 132
   bifurcation (*see* Bifurcation points)
   coincident, xvii, 10, 79, 86, 94, 106
   general (*see* General critical points)
   limit point (*see* Limit points)
   primary, 5, 11, 31, 134, 166
   simple (discrete), xvii, 3, 10, 11
   special (*see* Special critical points)
Critical surface, 111, 112, 134, 136, 137, 165, 168, 177

Index

Critical zone, 41, 111, 112, 125, 132, 133, 134, 147

D

DANIELSON, D. A., 26
Degrees of freedom, 3
  four, 195
  limited, 175
  one, 61, 64, 138, 148, 151
  two, 56, 67, 79, 203
Diagonalization, 7, 27, 113, 127, 160, 178, 181
  simultaneous, 179
Discrete systems, xvii, xix
Divergence, 209

E

Eigenvalue problems, 14
  linear, 109, 168
  multiple, 94
Eigenvector, 169, 170
Energy criterion, xvii, xx, xxi
Energy dissipation, xxii
Equilibrium equations, 4, 6, 8, 12, 13, 21, 27, 36, 127, 133
Equilibrium paths, 4, 6, 86
  complementary, 39, 40, 66
  fundamental, 7, 21, 27, 30, 32, 33, 36, 49, 51, 66, 69, 199
  post-buckling, 13, 14, 29, 30, 31, 32, 33, 45, 56, 66, 69, 78, 87, 88, 89
Equilibrium surface, 15, 38, 41, 96, 97, 111, 112, 115, 121, 123, 125, 127, 132, 147
  anticlastic, xvi, 38, 130, 131, 137, 151, 187
  degenerate (improper), 120, 121, 159
  fundamental, 124, 159, 160, 167, 173, 189, 203
  parabolic, 130, 131, 137
  post-buckling, 160, 161, 163, 165, 167, 190, 191, 193
  proper, 119, 121
  synclastic, 130, 131, 137, 166, 187
EULER, L., xv, 14, 26
EVAN-IVANOWSKI, R. M., 157
Exchange of stabilities, xvi, 30, 163, 167

Existence boundary, 111, 136, 137, 191 (see also Stability boundary)
Experimental results, 148, 154, 195, 206

F

Flutter, 209

G

Galerkin procedure, 3
General critical points, xvii, 115, 119, 121, 123, 127, 155, 173, 203, 205
  elliptic, 130, 134, 141, 157
  hyperbolic, 130, 134, 157
  parabolic, 130, 134, 157
  primary, 145
  singular, 119, 132, 141, 142, 144, 151, 155, 158
Generalized coordinates, xvi, 3, 4, 48
  critical, 11, 18, 80, 94, 127
  distinct set, 48, 50, 53
  principal, 8, 28, 86, 113, 160
  rotating, 28, 161
  sliding, 28, 161
GJELSVICK, A., 70
GODLEY, M. H. R., xviii
Gyroscopic systems, xviii, 209

H

HAYMAN, B., xviii
HENCKY, H., xvii
HO, D., 94
HUNT, G. W., 80
HUTCHINSON, J. W., xviii

I

Imperfection sensitivity, xvii, 36, 40, 41, 42, 45, 56, 98
Imperfections, xvi, xvii, 25, 26, 35, 37, 58, 60, 64, 77, 93
  parameter, 36, 69, 93

J

Jacobian, 87, 88
JOHN, K. C., 79, 92

Index

K

KAPPUS, R., xvii
KARMAN, T. von, xviii
Kirchhoff's uniqueness theorem, xv, 10
KOITER, W. T., xv, xvii, xviii, xix, xx, xxi, xxii, 33, 35, 79, 80

L

Lagrange's Theorem, xvii
LEIPHOLZ, H., 209
Lemma, 140, 141
Limit points, xvi, 7, 12, 25, 31, 61, 70, 97, 99, 100, 118
Linearly independent linear forms, 139
Load-deflection space, 4, 21, 49, 70, 110, 171, 177
Loading parameters, xvi, 3, 39, 103
   independent, xvi, 36, 109
   multiple, xvi, 109, 110
   single, xvi, 3, 109
Load-space, 110, 111, 144
LOO, T. C., 157
Lower bound, 158, 170
LYAPUNOV, M. A., xix, xx

M

MACLANE, S., 127
MARGUERRE, K., xvii
MASUR, E. F., 21, 47, 70
Multiple-parameter systems, xvi, 26, 35, 107, 109, 110, 123, 171

N

Nonconservative systems, xviii

O

One-way systems, xix

P

PAPKOVICH, P. F., 109, 157
PERRY, S. H., xviii

Perturbation, 50, 60
   independent parameters, 124, 126, 127, 183
   methods, 14
   multiple-parameter, 15, 26, 80, 113
   parameters, 15, 16, 18, 21, 24, 52, 55, 93
   pattern, 128
   single-parameter, 15, 43, 127
   two-parameter, 200
PLAUT, R., 170, 209
POINCARÉ, H., xvi, 167
Post-critical behaviour, xvii, 25, 56
Post-critical path (see Equilibrium paths)
Potential energy, xxi, 4, 5, 6, 21, 47, 48, 111, 113, 114, 123, 127
   first variation, 4
   second and higher order variations, 4, 5, 6, 18, 19, 81, 191
   surface, 17, 80, 81
Pseudo-conservative systems, 209

R

Rayleigh quotient, 170
Rayleigh-Ritz procedure, 3, 73
Regions
   existence, 111, 136, 137, 140
   inexistence, 111, 136, 137, 141
   instability, 141, 166, 167, 177, 188, 190, 191
   stability, 166, 167, 177, 188, 190, 191
REISSNER, H., xvii
RENTON, J. D., 109
ROORDA, J., 14, 26, 35, 37, 39, 70

S

SALET, G., 47
SEWELL, M. J., 15, 26, 79, 80, 88
SCHREYER, H. L., 21, 70
SOKOLNIKOFF, I. S., 9
SOUTHWELL, R. V., xvii
Special critical points, xvii, 119, 121, 123, 159, 173, 203, 205
   asymmetric, 163, 167
   primary, 165, 166
   symmetric, 181, 186, 188, 195
Stability
   boundary, 26, 109, 111, 123, 126, 136,

219

Index

Stability, boundary (*continued*)
   140, 141, 142, 156, 158, 164, 169, 171, 188
   coefficients, 10, 11, 25, 26
   coincident critical point, 80
   criteria xx, 146, 166, 167, 189, 190
   dynamical, xx
   statical, xx
   definition, xix, xx
   determinant, 5, 23, 31, 41, 53, 55, 125, 145, 165, 173, 188
   distribution, 31, 32, 33, 145, 166, 188
   equations, 24, 54, 55, 126, 174
   equilibrium state, 5, 6, 17
   simple critical state, 5, 6, 18, 31
Stable domain, 147, 166, 167
STEIN, M., 79
SUPPLE, W. J., 79
Symmetric systems, 47, 89, 159, 171, 181, 187
   potential energy, 47, 48, 89, 171
Symmetry, 47, 48
   individual, 89, 90, 92, 106
   properties, 182
   set, 89, 106

T

Taylor's series, 8, 14, 29, 32, 48, 113, 114
Theorem, 136, 137, 141, 142, 144, 147, 158, 165, 167, 170, 175, 176, 180, 191, 193

THOMPSON, J. M. T., xviii, 10, 15, 21, 25, 35, 69
TIMOSHENKO, S. P., 49
Transformation
   contravariant, 9
   covariant, 9
   orthogonal, 7, 27, 50, 68, 113, 127, 160, 161, 171, 178
   rotation, 150
   translation, 150, 178
TREFFTZ, E., xvii
Tri-furcation, 187, 188, 191

U

Upper bound, 158, 170

V

VAN DYKE, M. D., 14

W

WALKER, A. C., xviii
WESTERGAARD, H. M., 47
WOINOWSKY-KRIEGER, S. 49

Z

ZIEGLER, H., 209

QA
871
H9

SEP 23 1975